Articulating Novelty in Science and Art

Julian Stubbe

Articulating Novelty in Science and Art

The Comparative Technography of a Robotic Hand and a Media Art Installation

Julian Stubbe
Berlin, Germany

Unter dem Titel „Articulating Novelty in Science and Art. A Technography of Two Objects" zugelassene Dissertation an der Technischen Universität Berlin, 2016

OnlinePlus material on this book can be found on
http://www.springer.com/978-3-658-18979-2

ISBN 978-3-658-18978-5   ISBN 978-3-658-18979-2   (eBook)
DOI 10.1007/978-3-658-18979-2

Library of Congress Control Number: 2017946923

Springer VS
© Springer Fachmedien Wiesbaden GmbH 2017
This work is subject to copyright. All rights are reserved by the Publisher, whether the whole or part of the material is concerned, specifically the rights of translation, reprinting, reuse of illustrations, recitation, broadcasting, reproduction on microfilms or in any other physical way, and transmission or information storage and retrieval, electronic adaptation, computer software, or by similar or dissimilar methodology now known or hereafter developed.
The use of general descriptive names, registered names, trademarks, service marks, etc. in this publication does not imply, even in the absence of a specific statement, that such names are exempt from the relevant protective laws and regulations and therefore free for general use.
The publisher, the authors and the editors are safe to assume that the advice and information in this book are believed to be true and accurate at the date of publication. Neither the publisher nor the authors or the editors give a warranty, express or implied, with respect to the material contained herein or for any errors or omissions that may have been made. The publisher remains neutral with regard to jurisdictional claims in published maps and institutional affiliations.

Printed on acid-free paper

This Springer VS imprint is published by Springer Nature
The registered company is Springer Fachmedien Wiesbaden GmbH
The registered company address is: Abraham-Lincoln-Str. 46, 65189 Wiesbaden, Germany

Acknowledgments

I would like to begin my acknowledgements by thanking Raphael Deimel and Ralf Baecker, the two protagonists of the following story, who let me into their laboratory and studio so as to learn about their work and the significance of technological objects for epistemic and aesthetic practice. I wish them all the best for their careers and hope both read this account with interest as well as joy.

I am particularly indebted to my supervisor Werner Rammert for the inspiring discussions and the support he gave me throughout the project. His rejection of dogmatic theory as well as his critical deconstruction of technology in society without escaping into intellectual skepticism deeply inspired this study. I am indebted also to my second supervisor Estrid Sørensen, who provided valuable comments that undoubtedly improved my work.

I am thankful to the graduate school "Innovation Society Today: The Reflexive Creation of Novelty" at which this study was conducted. The four years of this study have been a wonderful and inspiring time especially due to our cordial atmosphere, the intense discussions at colloquia, workshops, and over lunch. I am particularly grateful to Dzifa Ametowobla, Jan-Peter Ferdinand, Miira Hill, Robert Jungmann, Uli Meyer, Jan-Hendrik Passoth, Fabian Schroth, Alexander Wentland, and Emily York, who read parts of my study and the related papers and provided rich comments and ideas.

Finally, this study would have never seen the light of day without Sabine and her continuous encouragement and emotional support, for which I thank her deeply.

Contents

PART I: Introduction, Study Design, and Theory

1 Novelty and Technological Objects 15
    1.1   Introducing the RBO Hand and *Mirage* 20
    1.1.1 The RBO Hand and the Development of Robotic Hands 21
    1.1.2 *Mirage* and the Advent of Technologies in Art 25
    1.2   Study Design I: Shaking off Taken-for-Granted Labels 30
    1.3   Theoretical Perspectives: Invention, Differential Pattern, or Biographical Passage 31
    1.3.1 Novelty as Invention 32
    1.3.2 Novelty as Differential Pattern 37
    1.3.3 Novelty as Biographical Passage 42
    1.4   Toward Novelty as Articulation 48
    1.4.1 Articulation 50
    1.4.2 Figures 54
    1.4.3 Technicity 58
    1.4.4 Enactment 62
    1.5   Study Design II: Abstraction, Critique, and Comparison as Method 64
    1.6   Overview of the Study 67

PART II: Analysis: Three Articulations of Novelty

2 Identity: How Loose Elements are Connected 73
    2.1   Enacting the Past through Stories and Their Tangible Traces 75
    2.1.1 Stories of Deviation 75

2.1.2 Material Traces 78
2.2 Enacting Potentials through Figures, Prototypes, and Bodies 83
2.2.1 Ideas and Their Figures 83
2.2.2 Embodying Material Potentials 88
2.2.3 The Bodily Rendering of Future Objects 90
2.3 Articulating an Object Identity 95
2.3.1 Four Articulations of Ideas 95
2.3.2 Object Identities 97
2.3.3 Novelty as Object Identity 104

3 Form: How Materials Become Effective 107
3.1 Formats and Their Technical Features 111
3.1.1 Technical Features of Robotic Hands 112
3.1.2 Technical Features of Cybernetic Machines 115
3.1.3 Formats and Their Core Functionalities 118
3.2 Hybrid Constellations, Inquiries, and Distributed Agency 119
3.2.1 Robotics Infrastructure and "Everyday Life" in the Laboratory 120
3.2.2 Anticipated Conditions and Inquiries with Open Hardware in the Studio 124
3.2.3 Distributed Agency 128
3.3 Assembling Technical Forms 130
3.3.1 Embodiment as Epistemic and Artistic Stance 131
3.3.2 Materials and Manufacturing Method 134
3.3.3 Kinematic Architectures 139
3.3.4 Actuation 141
3.3.5 Sensors and Input Signals 144
3.3.6 Converters 146
3.3.7 Outputs 148
3.4 Articulating Novelty through Technical Forms 154
3.5 Excursus: Two Techno-Aesthetics 158

4	Difference: How Categories are Valorized	161
4.1	Discursive Practices	166
4.1.1	Conferences, Festivals, and Galleries	166
4.1.2	Journals, Books, and Catalogues	168
4.1.3	Graphs, Models, and Technical Drawings	168
4.1.4	Images, Videos, and Visual Codes	169
4.2	Translating Events into Other Entities	173
4.2.1	The RBO Hand: Translating Events into Charts and Pictures	173
4.2.2	*Mirage:* Translating events into Drawings, Pictures, and Video	178
4.2.3	Stabilizing, Concealing, and Travelling of Relations	183
4.3	Referencing by the Authors	184
4.3.1	Referencing the RBO Hand	185
4.3.2	Referencing *Mirage*	192
4.3.3	Conceptual, Figurative, and Associative References	197
4.4	Referencing by the Archive	198
4.4.1	The RBO Hand in Citations, Awards, and Online Newspapers	199
4.4.2	*Mirage* in its Exhibition, Online Articles, and Prizes	206
4.4.3	Passage Points, Selected Elements, and Dynamic Practices	213
4.5	Articulating Difference as Novelty	214

PART III: Discussion and Conclusions

5	The Aesthetic Reflexivity of Material Practice	221
5.1	Which Reflexivity?	222
5.2	The Indexicality of Referencing	224
5.3	Aesthetic Reflexivity and Materiality	226
5.4	From Symbols to Allegories of Technology	228

6	Articulating Novelty	233
6.1	Individuating and Relating Objects through Identity, Form, and Difference	234

	6.1.1	Novelty as Individuation and Relation 234
	6.1.2	Identity 235
	6.1.3	Form 237
	6.1.4	Difference 239
6.2		Articulations: (Differently) Connected Figures, Technicity, and Enactments 242
6.3		Outlook 246

References 249

The appendix to this book is freely available to download at the author's product site at springer.com.

List of Figures

Figure 1:	The RBO Hand.	22
Figure 2:	The Salisbury Hand.	23
Figure 3:	*Mirage.*	26
Figure 4:	Gordon Pask's *The Colloquy of Mobiles*	28
Figure 5:	Hand-like shape in the laboratory's workshop.	80
Figure 6:	Structure for testing the translation of the electric signal into a mechanical pull.	81
Figure 7:	Starfish Grabber.	89
Figure 8:	Enacting the silicone's softness and potential for grasping.	91
Figure 9:	Enacting the "landscape situation."	93
Figure 10:	The SDM Hand.	113
Figure 11:	Ross Ashby's Homeostat	117
Figure 12:	Inquiry patterns in the RBO Laboratory.	123
Figure 13:	Baecker using his body as actuator and epistemic tool.	126
Figure 14:	Test structure made from wooden plates, hooks, and strings.	127
Figure 15:	Test structure made from Arduino board, bows of acrylic glass, and wires.	127
Figure 16:	The RBO Hand's technical form.	131
Figure 17:	*Mirage's* technical form in Baecker's studio, close to its first exhibition.	132
Figure 18:	The composition of the PneuFlex Actuator.	136
Figure 19:	Early experimental setup of cross-line laser and mirror.	139
Figure 20:	Skeleton made of acrylic glass.	144
Figure 21:	Computer screen showing frequently changing signals and their algorithmic variation.	147
Figure 22:	The RBO Hand performing a surface-constrained grasp.	150
Figure 23:	*Mirage's* moving image.	152
Figure 24:	Shadow Hand holding a light bulb.	171
Figure 25:	Two kinds of failure in an experiment with the RBO Hand.	175
Figure 26:	Graph translated from grasping experiment.	178
Figure 27:	Technical drawing of *Mirage*'s design approximately six months prior to the exhibition.	180

Figure 28:	Video-screening of human hand grasping a sponge at the ICRA.	189
Figure 29:	Profile of the Alps from Baecker's Twitter account.	196
Figure 30:	Screenshot of the *Spiegel Online* article on the RBO Hand.	205

PART I
Introduction, Study Design, and Theory

1 Novelty and Technological Objects

"La genèse de l'objet technique fait partie de son être."
Gilbert Simondon (1958)

This study is about novelty and technological objects. It is about the unfolding and revealing of what was not there before, and how the new individuates and enters a technical form. Technology is an ambiguous term in this regard: on the one hand, stability, repetition, and predictability characterize technology, and, on the other hand, technology challenges order, as it is poietic. The striving for stability reflects the former meaning of technology as a stabilized set of relations that repetitively produces expected results. If one applies technology, one can expect certainty, predictability, and repeatability at the level of epistemic as well as social practice. Technology involves the fixing of relations, in this regard. In contrast, technology also implies assembling and arranging new relations, connecting things that were not connected before. Martin Heidegger spoke in this regard of technology as a revealing (*Entbergen*), which addresses a moving into presence and an unconcealment of relations (Heidegger 1977). This expresses technology's potential to form, create, and articulate new relations among previously disconnected entities. Technology is not only a matter of stabilizing existing relations then, but also a matter of contending these. In this respect, novelty seems like a natural companion to technology, as technical forms imply assembling and re-arranging elements. This paradox between stabilizing and revealing relations signifies technology and the different forms it enters.

The study takes its beginning in a simple but intriguing empirical similarity concerning technology's latter meaning: the poietic engagement with technology in science as well as in art. At first sight, science seems to be the more natural habitat for technology, and technology in art occurs more as a contemporary phenomenon. Big science dominates the public image of science, with its high-technology infrastructures and the promises of technological progress and innovation that go along with it. The scientific mode of revealing the world and the institutionalization of experimental science in the 17th century deeply connect with the development of epistemic hardware like technical apparatuses, instruments, and experimental settings, which have found their temporary climax in superstructures like CERN's particle accelerator. However, it was never big science alone that was concerned with technology. There is a certain tinkering culture in science, too, which fosters small-scale projects of developing and engineering technological objects (cf. Pickering 1995). This is, for instance, how gravity waves were rendered first by a community of scientists who engineered their own experimental settings (Collins 1982), or how cyberneticists build technical apparatuses instead of only running abstract simulations (Pickering 2002). In those cases, the engineering of hardware and the physical creation of behavior occur as an epistemic counterpart to the rather clean grammar of abstract theory, as well as to the top-down development of technical infrastructure through high-finance projects.

Technology is certainly not alien to art if one understands technology in a broad sense. The Greek term *téchne* already resonates with the quest for stabilized methods and rules in the production of art. The rules structure for producing art are a success story in and of itself and is translated into "creativity techniques" with various aesthetic applications in postmodern societies (Reckwitz 2012). However, this is not what I mean when referring to the similarity of poietic engagement with technology in science and art. The similarity refers to technology in a much more narrow sense and, in particular, to the advent of technologically complex artworks as in media art, which comment on technological development, and wherein the constructed objects require sophisticated skills of electrical engineering and programming. In this sense, technology appears somewhat as a contemporary phenomenon in contrast to technical traditions in science. The advent of artworks that implement the latest technologies began in the 1960s, when artists like Nam June Paik and Edward Ihnatowicz began building cybernetic sculptures equipped with sensors, electronics, microchips, and other technologies that required engineering skills and not only expressive methods. For instance, Ihnatowicz's *The Senster* from 1968-70 is a movable sculpture that reacts to movement and sound from the audience. Ihnatowicz implemented state-of-the-art sensory and microchip technologies from electronics developer Philips for its technical construction (Zivanovic 2005). This kind of technological engagement is not a singular case in media art, and there has already been a certain tradition for artworks that implement the latest technologies. However, literature on media art is mainly concerned with documenting objects and discussing their symbolic and discursive meanings. This is no surprise, as media art is almost exclusively the subject matter of the history of art and media studies and has found little recognition in science and technology studies or the sociology of technology. This narrow focus on symbolic meaning conceals the material dynamics of technological engineering. In contrast, in the present study, I approach media art with a focus on technology and from the perspective of science and technology studies.

Similarly to the ethnographic approach to studying science, which began moving into the laboratories of hard science in the 1980s so to render visible the technological breeding space of scientific knowledge, I move into a robotics laboratory as well as into a media artist's studio. In this sense, I approach the initiating observation by comparing cases from both arenas through a shared lens. This *technographic* lens blends heuristic resources from science and technology studies and focuses on the technological aspects of scientific and artistic practice. In this sense, it focuses on the micro-scale of engaging with technology. The science object of that comparison is a robotic hand made from silicone. It is called the RBO Hand. Silicone is not a common material used for robotic hands. Usually, robotic hands are made from solid materials that are electronically steered. Hence, the silicone hand's grasping is not programmed, but based on compliance with an artifact's surface. The hand is a current research challenge for a robotics laboratory. The art object of the comparison is a media installation called *Mirage*. For its investigation, I visited an artist's studio and accompanied him during his creative process. The installation consists of different mechanical and electronic elements assembled to create a kind of floating movement. The movement is visualized through a laser projection that reacts to the contingent interaction of these elements. Both objects are technically complex; that is, their engineering requires advanced technical knowledge and a specific material infrastructure to build them. Furthermore, their complexity entails opacity, which renders the exact principles of their inner workings hidden not only from the lay spectator, but also from all actors involved. Over the course of approximately two years, I encountered both objects' developments through various situations. I visited experiments in the laboratory and studio, followed mundane tinkering practices, recurrently conducted interviews, went to robotics conferences and art exhibitions, and analyzed both objects' international discursive recognition and valorization through citation, awards, and documentation.

The focus of this study is not on the role of technology in practices as such. It does not examine how technology infiltrates epistemic and artistic practice, in the sense of stabilizing a given set of methods or creative techniques. Surely, how technology supports practices plays a role for empirically inquiring into scientific and aesthetic conduct, but the main issue of the study is the *object*-character of technology. An object, in this regard, is something produced and perceived as exterior to the self; that is, opposing and reflecting human activity. It is neither the material artifact nor its symbolic meaning; it is the sense of coherence that runs through both geneses. The object is not a given, simply there, nor does it belong, by nature, to a specific kind. Rather, it is produced as something individual through its consistent and converging qualities. Philosopher of technology Gilbert Simondon expresses this, saying, "the technical object is unity as a unity of becoming" (Simondon [1958] 2012a, 20). It does not precede its becoming, but the object is the sense of unity and coherence that is present in every state of becoming. It is not a fixed materialized unit, but a composition that requires rendering to exist as an entity.

The aim of this study is to characterize *novelty* within the different states and compositions of an object's becoming. It is not concerned with the social meaning of novelty – what actors understand as new, progressive, or wishful. These questions are already tackled by the larger cannon of sociological innovation studies as well as by critical science and technology studies, which both discuss how technology and innovation embodies societal desires for progress and salvation (i.e. Pinch and Bijker 1987; Giddens 1990; Braun-Thürmann 2005; Hutter et al. 2015). In contrast, this study aims at saying something more substantial about novelty. It attempts to delineate certain shifts or thresholds in an object's becoming that individuate an object and render its difference visible. This is not to say that novelty is a stable property of an object. On the contrary, it might change throughout an object's genesis: novelty might appear as a reassemblage or as a rupture in the flow of things, as potential or as efficacy. This entails attending the enactment of novelty through cultural imaginaries and narratives about technologies, as well as acknowledging the shifts in technical forms that make loose elements enter a new kind of circularity. The main question for this study is: *what is novelty in the becoming of technological objects, and how does it become part of a shared reality?*

Drawing on the body of existing literature, the study distinguishes three established theoretical perspectives on novelty and technological objects: novelty as invention (i.e. Gilfillan 1952; Schumpeter 1939; Hughes 1987), as differential pattern (i.e. Latour and Woolgar 1986; Rheinberger 1992; Pickering 1995), and as biographical passage (i.e. Kopytoff 1986; Groys 1997; Daston 2000). Despite certain points of critique, the aim of the present study is to *render visible* an issue that these perspectives only *implicitly* carry: the tension between individuating and relating that characterizes an object's novelty. Individuation makes an object one, and relating makes it one of the many in order to mark its difference. This central tension characterizes novelty, whose different forms this study seeks to render visible. It stresses that novelty is not only difference as such, but that difference requires coherence among the diverse elements of an object's compositions to become novelty.

To cope with this analytical challenges, I propose *articulations* as heuristic. Articulations are the sense of unity that appears through connecting diverse elements. They are the point of convergence where loose elements click in and unite. In this sense, articulations stress that there is no natural belongingness of elements and coherence is an accomplishment that can take different forms (Hall 1986; Slack and Wise 2005). Connections can always be different, but coherence makes loose elements into an object that is distinguishably different from others. The diversity of elements is central to articulations: articulations consider material elements to be selected, fitted, and connected to chains of efficacy (Latour 1999), as well as semiotic categories to be stabilized through collecting attributions of value, desire, or promise (Haraway 1989). Diverse things are articulated, when their difference is not contradicting, but when differing characteristics are aligned and fitted so as to appear and behave as a distinct entity. Articulations are different from expressions, as they are not concerned with whether or not a wording matches a state of affairs. Instead, articulations stress how connected elements resonate, click in, and create sensible efficacy.

Considering articulations as a comparative heuristic allows the pre-setting of certain theoretical elements that are of significance for the central concerns of the study. Such elements can serve as comparative issues that allow one to draw connections between empirical cases as well as move onwards and learn about theoretical issues concerning technology and novelty (cf. Strathern 1991). These elements are figures, technicity, and enactments. This is the central triad of this study, as they are the basic elements that constitute the articulation of technological objects. *Figures* capture the semiotics of stories, imaginaries, and discourses that narrate the significance of technology (Haraway 1989; Suchman 2007). *Technicity* is the technical character that pulls in technical elements to form causal circularity (Simondon [1958] 2012a). *Enactments* are the activities of acting out an object's character and putting its material capacities into effect (Rammert 1999; Barad 2007). I regard these three elements as being connected and as continuously resonating with how an object becomes distinguishable. However, the central issue is that they connect differently depending on an object's location and temporal state. It is the different articulation of these elements that I seek to delineate in this study as articulations of novelty.

In the study's analytical Part II, I infer three articulations by comparing the becoming of the RBO Hand and *Mirage*: identity, form, and difference. Each of these articulations emphasizes another element. *Identity* is concerned with how coherence evolves before an object is technically realized. The articulation connects histories about the RBO Hand's and *Mirage*'s origins, figures of robotics and artificial intelligence, as well as prototypes and the actor's bodily performance of the future object's characteristics. *Form*, in contrast, emphasizes how both objects concretize their technical characters. The technical form connects the diverse agents of a technical constellation so as to articulate a distinct kind of physical efficacy. *Difference* shifts the focus onto the discursive valorization of the RBO Hand and *Mirage*. Regarding difference as an articulation places emphasis on how selected categories increasingly stabilize as embodiments of an object's novelty. These three articulations are conceptual aggregations and not empirical categories. Each of them tackles a conceptual issue, which allows intellectual movement away from the specifics of fields and towards better understanding the tension between individuation and relation in different states of an object's becoming.

The closing Part III of this study begins by jumping back to the initial "simple, but intriguing similarity." Based on the empirical analysis in the preceding chapters, it asks if the technological engagement in science and art and, in particular, the material tinkering entailed in the RBO Hand's and *Mirage*'s production are not so simple or such singular instances after all. Indeed, a turn toward creating technical behavior and efficacy, which signify the concepts behind both objects, might well be understood as an opposing stance that counteracts a cognitivist and "emptied-out" (Giddens 1990) kind of abstract theory. This turn toward materiality, effect, and experience signifies a critical engagement with technology, science, society, and their institutionalized modes of representation. The aesthetic reflexivity of engaging science, art, and technology is not a singular instance, but can be identified in other contemporary science and art projects, too (cf. Lash 1993; Lash and Urry 1994). In this sense, the RBO Hand and *Mirage* are examples of aesthetic reflexivity in the diverse articulations of technology, science, and art.

In the remainder of Part I, I will first continue introducing the RBO Hand and *Mirage*. Sketching the sampling of both objects entails a prompt to shake of such taken-for-granted labels as "science" and "art" in favor of being open to how differently both are articulated in the objects' becoming. Three theoretical perspectives for understanding the relation between novelty and technology follow the empirical introduction: invention, differential pattern, and biographical passage. These perspectives allow the characterization of the relation of novelty and technology without falling back on arguments that explain novelty and technology as field-specific modes of production. Hence, all three perspectives are inevitably abstract so as to allow comparison of novelty and technology in diverse societal fields. By sketching each perspective's shortcomings, I propose understanding novelty as articulation. I sketch the term's basic implications before moving on to its three basic elements: figures, technicity, and enactment. Discussing what this heuristic implicates, and what role abstraction and critique play, also entails re-stating the general question above: *how do technological objects articulate novelty?* This is the central question of this comparative study. In closing the introduction, I make some methodological remarks on comparison and, finally, give an overview of the following chapters.

1.1 Introducing the RBO Hand and *Mirage*

The RBO Hand and *Mirage* are both complex technological objects. They are not historical monoliths, but developed within specific social and institutional settings as well as in a certain historical context. The following introductory remarks outline these settings and sketch a brief historical trajectory of robotic hands and media art installations in general.

1.1.1 The RBO Hand and the Development of Robotic Hands

RBO stands for "Robotics and Biology Laboratory," which is the name of the institute of the Technical University Berlin, where the Hand[1] is being developed and engineered. The RBO Hand is a current research project at the institute, led by the director, Professor Dr. Oliver Brock. The institute's research is not limited to robotic hands, but is also engaged with several domains of humanoid robotics. Its research is basic, as it is less concerned with concrete applications of developed technologies and more with fundamental processes of robotic behavior. In this regard, the RBO Hand is a contribution to the research field of robotic grasping, where basic concepts and approaches to the design of robotic hands are being discussed. The development of the RBO Hand fits into the general framework of the institute, which is concerned with robotic perception, learning, and manipulation based on a robot's interaction with the environment. In this sense, there is a connecting thread between different research projects at the institute, whose focus is on environmental interaction. Robotics researcher Raphael Deimel mainly conducts the development and engineering of the RBO Hand.

The Hand's central characteristic is its material – silicone (*Figure 1*). It does not have a silicone skin in the sense of an anthropomorphic soft cover. Rather, its complete mechanical body consists of silicone, which implies its actuators as well as a contact surface. Silicone is not a common material used for robotic hands. Usually, robotic hands are made from solid materials that are electronically steered. Instead, the specific design of the Hand attempts to exploit the softness and deformability of silicone for robotic grasping. Hence, the Hand's grasping is not programmed or steered with sensors, but based on compliance with an item's surface. Throughout this study, the RBO Hand appears in very different material shapes – from a prototypical gripper design to a sophisticated hand-like form. When readily assembled, the RBO Hand consists of three silicone fingers and a two-part silicone palm. Each finger has an internal air channel, which is inflated so as to make it bend. Due to their two-layer design, comprised of one deformable and one resistant layer, the fingers bend in a directed way so as to form a closed cage that may capture an item. For grasping experiments, the RBO Hand is mounted on a standard robotic arm, which drives the Hand into an optimal position for grasping. The opportunity to connect the Hand to a larger standardized infrastructure allows its development as a standalone research project.

---

1   I will capitalize the term "hand" throughout this study when I refer to the particular RBO Hand without adding a signifying prefix.

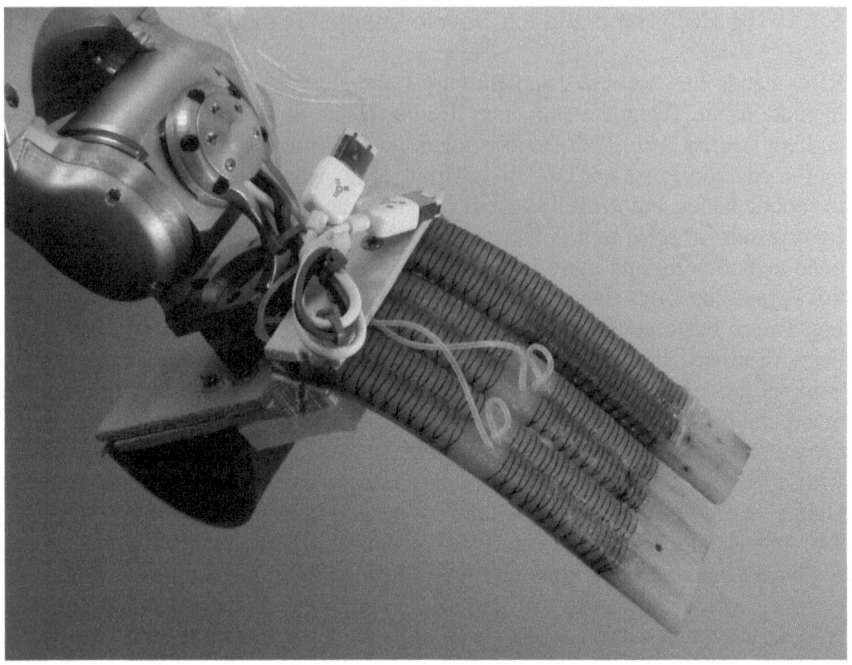

Figure 1: The RBO Hand (source, Deimel and Brock 2013).

At first glance, the RBO Hand's design is distinctively different from what is commonly expected from a robotic hand. The timeline of modern robotic hands begins in the 1950s with teleoperated manipulators (Rosheim 1994). A person remotely controlled these manipulators, which led to their nickname "master-slave systems." The development of these grasping machines was driven by their potential application in industrial settings and as part of technical infrastructures in scientific laboratories. Their development received a prominent position and governmental support, as they were often installed to handle radioactive material. Their master-slave design enabled the laboratory staff to stay away from radiation and use the grippers to handle risky substances. The common gripper design was based on two parallel jaws that allowed basic grasping functions. For the development of industrial robots, this design was significant, as it allowed for dexterous grasping, given the environment was structured and predictable. The two-jaw gripper design was stabilized early and is still common among today's industrial grippers. However, these grippers might perform certain dedicated tasks well, but fall short in general-purpose applications.

1.1 Introducing the RBO Hand and *Mirage*

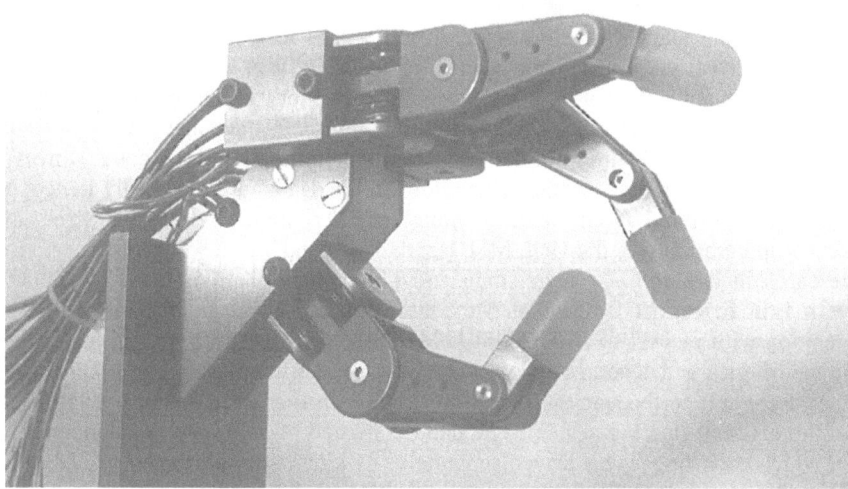

Figure 2: The Salisbury Hand (source, Rosheim 1994).

An important reference in the history of robotic hands is the Stanford/JLP Hand, developed by Kenneth Salisbury in the 1980s (*Figure 2*). Despite its general significance for robotic grasping, its development entailed aspects that are particularly important concerning the RBO Hand. In the Springer Handbook of Robotics, the "Salisbury Hand" is marked as the first robotic hand designed for dexterous manipulation (Prattichizzo and Trinkle 2008, 671). It allows for a far broader range of movement than two-jaw gripper designs and does not rely on teleoperation, as it is equipped with sensors. The hand has three-jointed fingers, each with three degrees of freedom. Each finger is connected with four Teflon coated cables that run through special conduits. To reduce the number of components, all three fingers are designed modularly. Finger positions are controlled by information generated by strain gauge sensors and motor position sensors located behind each proximal joint. The fundamental grasp modeling and analysis done by Salisbury still provides a basis for the grasp synthesis and dexterous manipulation research that continues today. Besides its technical superiority compared to previous grippers, the Salisbury Hand has further significance as it is a standalone robotic hand. Thus, it marks a development towards a research field focused on grasping without advancing a complete robotic system – robotic hands have become modular. Furthermore, the Salisbury Hand is able to perform a wide range of different dexterous grasping without a strictly anthropomorphic design.

There are a number of other hands in the young history of robotic hands, each of which mark a specific point of progress as well as the limits of certain design approaches. The Hitachi Hand, for instance, was a stepping-stone, as it used the advantages of a new material: shape-memory alloy. The material "remembers" its original shape and returns to its pre-deformed shape when heated, which is a principal similar to the RBO Hand's exploitation of silicone. However, the Hitachi Hand had limits. Like all metal hands that are repeatedly flexed, it eventually suffered fatigue and grasping failed. A pioneer design in terms of sensor integration was the Utah/MIT Hand. This hand was developed to perform research in laboratories on grasping and finger manipulation. It was equipped with Hall Effect sensors, which vary their output voltage in response to their position with respect to a magnetic field. These provided information about the joints' angles and tendon tension. Positioned in the hand's wrist and knuckles, they were able to feedback information about each finger's position with respect to the external steering system. The hand consisted of a complex tendon system in which each finger was driven separately. Its kinematics consisted of 288 pulleys, making its calibration a complex task. However, the MIT Hand is supposed to have dead-ended because of its overly close emulation of nature (Rosheim 1994, 225).

In modern robotics, anthropomorphic design is a contested paradigm. When robotic hands cooperate with humans or when they are teleoperated by humans like prosthetic hands, anthropomorphism is a common design objective. In robotics discourse, another design approach is a *minimalist* approach, which demonstrates a preference for the simplest mechanical structure, the minimum number of actuators, the simplest set of sensors, etc. that fulfill the task requirements. (Controzzi, Cipriani, and Carrozza 2014, 230). For instance, the Salisbury Hand implements the minimalist approach – it has just three fingers with sensors on the tips. A modern example of a minimalist design is the SDM (Shape Deposition Manufacturing) polymer hand, which has been developed as a novel adaptive and compliant grasper that can grasp objects spanning a wide range of sizes, shapes, masses, and positions using only a single actuator (Dollar and Howe 2010).

This short introduction to robotic hands illustrates that the RBO Hand and its silicone-based design is distinct from most robotic hands, but also builds upon pre-existing design approaches. Its development is significantly driven by its uncommon material, which affords a specific design of fingers, palm, and additional elements to be used for robotic grasping. Finding the functioning allocation of elements and concretizing the material components into a working unit is in the focus of the RBO Hand's development. Deimel's design activities and accounts of his research practice focus on the realization of technical relations that work and behave in a certain way, rather than being concerned with the Hand's outward appearance. Hence, the development of the RBO Hand is significantly structured by the silicone's material characteristics, as there is no blueprint design that can be used as a template. Surely, the focus on technical relations in the Hand's design alone does not constitute its novelty. In this study, I investigate how the RBO Hand's existence entails the articulation of symbols, bodies, and materialities. Nevertheless, the focus on technology is significant for its sampling, as it allows investigation of how previously unrelated elements gain coherence, something object-like. This focus on material efficacy, which I observed during my first visits to the RBO Laboratory, also informed the sampling of my second case.

1.1.2 *Mirage* and the Advent of Technologies in Art

My second case is a media installation by the artist Ralf Baecker, called *Mirage*. Baecker is a freelance artist based in Berlin with an educational background in computer science. In general, his art is concerned with technology as a medium as well as a topic. According to his own description, Baecker builds "systems and machines which explore the poetic potential of technology."[2] Most of his artworks are concerned with a kind of "technological life on its own" and, in this regard, attempt to render commonly hidden technological processes visible. Thereby, he combines digital, electronic, and mechanical elements into complex assemblages that behave in a contingent but technically directed manner. Several of his works have been internationally recognized, discussed, and awarded by media art discourse.

For *Mirage*, Baecker assembled different mechanical and electronic elements to create a kind of floating movement projected on a wall (*Figure 3*). In its finished state, the installation is equipped with a sensor that registers the magnetic field of the Earth. The sensor is sensitive enough to track continuous minor changes of the Earth's magnetic field that are dependent on the Earth's geodynamics and their interactions with the sun, but remain undetected by human perception. The sensor's signals feed into a complex, unsupervised learning algorithm that calculates a prospective trajectory of the signal. The trajectories are beyond deterministic human control, as they are based on randomly occurring signal patterns and do not orientate their learning based on pre-defined system values. These digital variations translate into analogue signals that actuate 48 muscle wires connected to a thin and flexible mirror sheet. In effect, the surface of the mirror sheet changes according to the translated signals. The continuous movement of the mirror sheet is rendered visible through a thin, horizontal cross-line laser that is directed onto the mirror surface. Through its flat angle of

---

2   Baecker's personal website: http://www.rlfbckr.org/ (last accessed 31st August 2015).

incidence, the laser touches the mirror sheet slightly and reflects a contingently moving image onto the opposing wall.

Figure 3:   *Mirage* (source, Ralf Baecker 2014).

Baecker's descriptions and lectures concerning his artworks often entail references to cybernetic machines. The self-regulation and ecological composition of philosophical machines constructed by early cybernaticians, such as Ross Ashby's *Homeostat*, are issues that Baecker reiterates through several of his artworks. The technical composition of *Mirage* also entails principals of cybernetics, such as its environmental adaptation. Connecting principles of cybernetics and art is not a novel approach as such and had prominent historical forerunners in the 1950s and 60s, such as, for instance, Nicolas Schöffer, Gordon Pask, and Roy Ascott. The exhibition "Cybernetic Serendipity – The Computer and the Arts," which took place in London in 1968, was a key event for new artistic forms that dealt with issues concerning new technologies as well as used technologies as a medium. As an example, Pask exhibited an installation called *The Colloquy of Mobiles* (*Figure 4*), which was a computer-based, reactive learning system of five mobiles that were hung from the ceiling and reacted to each other via changes of light and sound (Reichardt 1968, 34-5). Other installations were concerned with computer graphics or sound compositions that rendered algorithmic patterns and gave form to the calculating processes inside, at the time, such novel and sublime machines. Whereas Pask approached art from a scientific theory-driven perspective, Ascott appropriated cybernetic principles as an artist. Ascott was attracted to cybernetics' general world view, which implied inquiries into the dynamic and contingent processes of information transfer amongst machines, flora, fauna, and humans and the alteration of behavior at the system level (Shanken 2002). In 1968, he proposed a cybernetic stance on art and articulated an emphasis on ambiguity, mutability, feedback, and behavior as paradigmatic foci of his behaviorist art program (Ascott 1968). Technology played a vital, but not superior, role in realizing this cybernetic vision. Ascott regarded technology as a means to enhance human creativity at the individual level, as well as to enable collaborative interaction between participants from diverse fields (cf. Shanken 2002). In this regard, the artistic appropriation of cybernetic principles has a history that implies engagement with technological objects as a topic and medium for art. Baecker's artistic technologies re-discover the cybernetic principles as the conversion of signals, contingency, and adaptation. However, this does not imply that Baecker follows an artistic program like that proposed by Ascott. Rather, he shares an interest in the dynamics and contingencies of systems as addressed by early cybernaticians and now reiterates these in his art objects, which combine digital, analogue, and mechanical technologies as both topic and medium.

The larger umbrella category of Baecker's artworks is media art, which is less programmatic than cybernetic art and is comprised of artworks that use various materials, such as computer graphics, animations, light projections, sound, robotics hardware, etc., that are not regarded as traditional art forms. A large share of media artworks not only use new media as expressive means, but are concerned with issues of new media usage and human-technology relations. Hence, a self-referential use of media is a signifier of media art that co-develops with the societal use of technology (Fleischmann and Strauss 2007; Zielinski 2011). This rather broad collective topic shared by media artworks makes trajectories in the field difficult, as they could begin in several different related domains, such as experimental film, kinetic art, or early computer art. Nevertheless, the issues that Baecker addresses through his artworks are within the scope of the thematic range of media art because they are concerned with self-referential renderings of technological processes; that is *not* to say, they are *typical* media artworks, as they are not much concerned with human-media relations.

Figure 4: Gordon Pask's *The Colloquy of Mobiles* (source, *Cybernetic Serendipity*, ICA London, 1968).

Aside from this thematic labeling of Baecker's art, there are certain artworks that share some of *Mirage*'s technical elements. For instance, *The Senester*, by Ihnatowicz, is a robotic sculpture equipped with sensors, similarly to *Mirage*, in order to react to changes in its environment. The sculpture is large – approximately 5 x 2.4 meters – and has an animal-like appearance. It is equipped with microphones that allow the sculpture to move its head in the direction of a sound source. Furthermore, it has two Doppler radar units, which sense people's movements and, hence, allow immediate interaction. For the design and engineering, Ihnatowicz was able to use technical resources from Philips' development laboratories, which made *The Senester* not only a sophisticated piece of art, but also a state-of-the-art technological object (Zivanovic 2005). Several media artists have already used laser technology, which is another of *Mirage*'s main technical elements. Among them is media art pioneer Nam Jun Paik, who used laser projection for a spatial installation called *Baroque Laser*, which he installed in a baroque church near Münster, Germany in 1995. The installation combined laser and video technologies, whereas much of the sculptural quality of laser projection was achieved through its maneuverability via mirrors, enabling a "kaleidoscopic orchestration of the space."[3] Ihnatowicz's and Paik's appropriation of new technologies is not atypical for pioneering media art, which often entails experimental implementations of state-of-the-art technologies.

In this regard, Baecker's artworks and *Mirage* in particular combine different topics and technologies that are not mainstream in media art, but that still have certain forerunners. Baecker's topic is the hidden processes that make technologies work, but which are concealed in everyday life practice, like algorithms or the self-adaptability of technical systems. Similarly to Ihnatowicz and Paik, this topic makes the usage of sophisticated technologies a natural choice, and implies much experimenting and tinkering in Baecker's artistic practice so as to integrate algorithms, sensors, mechanical elements, etc. into functioning technical assemblages. As I observed in early visits to his studio, his work is much concerned with the entailed and simple problem: making things work and creating a technically functioning unit. In this regard, Baecker's artworks not only need to work as artworks, but they need to work as technological objects. For *Mirage*, this means the artwork is not only the laser image produced, but also the way the image is generated as a product of contingently adaptive technical relations.

---

3   Rudolf Frieling's description of Paik's artwork on MediaArtNet: http://www.medienkunst netz.de/werke/baroque-laser/ (last accessed September 3, 2015).

1.2 Study Design I: Shaking off Taken-for-Granted Labels

In order to introduce the RBO Hand and *Mirage*, I did something typical for studying the novelty of a particular object: I sketched how much both objects relate to and differ from other objects of that particular category. In these cases, such taken-for-granted categories are (among others) "scientific object" and "art object." By fixating on these categories *a priori*, one can study which elements are reiterating pre-existing issues and what goes beyond the state-of-the-art of a particular field. However, for this study, I will no longer follow such argumentation. It is not my aim to discuss whether the RBO Hand means true progress for robotic grasping, or if *Mirage* embodies a new art form. I will leave this discussion to the actors in the field and their discourses. Instead, I want to begin this study by shaking off the most obvious labels of "science" and "art" and focus on the similarities of both cases, which is their distinctive concern with materiality as part of technically complex constellations. Focusing on similarities allows one to study how their technical characters shape the geneses of both objects and, furthermore, how novelty is a temporal product that might occur very differently across varying situations but that shares similarities across presumably rather distant cases. This is not to say that history does not matter. On the contrary, it means opening up the study design to how differently material and semiotic resources are enacted and what parts of particular histories are selected so as to give meaning to an object. Labels like "science" and "art" are not *a priori* categories in this regard, but are actively produced distinctions. How distinctions and unity are produced is the subject matter of this study.

But what do I mean by stressing the similarity of both objects' geneses within technically complex constellations? It means their engineering requires advanced technical knowledge and a specific material infrastructure to build them. In both cases, technical knowledge is required, for instance, for programming and electrical engineering – not only to build the technical body of both objects, but also to work with the infrastructural setting of both objects. These settings include scientific robotics hardware as well as digital technologies for operating hardware components in the studio. Furthermore, both objects' complexity entails opacity, which makes the exact principles of their inner workings not only hidden from the lay spectator, but from all actors involved. Surely, the RBO Hand is a minimalist approach to robotic design, and its hardware components are also very visible, but this does not mean the Hand's behavior can be modeled in advance. On the contrary, the characteristics of the silicone exceed precise control, and their integration requires explorative tinkering that orientates towards behavior and not modeling. For *Mirage*, this is similar since the assemblage of its plentiful hardware components requires explorative tinkering practices, over the course of which complexity and contingent behavior increases. These characteristics foster what I have already mentioned: the epistemic and artistic practices focus on the technical character of both objects and realization of a technically functioning unit. These similarities have informed this sampling and made the RBO Hand and *Mirage* sensible choices for studying the relation of novelty and technological objects.

Furthermore, choosing a comparative study design of only two cases allows analysis of novelty in differing modes – that is, how novelty is conveyed differently depending on an object's temporal state and the situation of its enactment. Hence, over the course of approximately two years, I encountered both objects' developments through various situations. In a broad sense, I created ethnographies with additional sources: I visited experiments in the studio and laboratory, followed mundane tinkering practices, recurrently conducted interviews, went to robotics conferences and art exhibitions, and analyzed how the RBO Hand and *Mirage* have been cited, awarded, documented, and discussed by international robotics and media art discourses.[4] Nevertheless, this study does not describe a trajectory of both objects' design phases from first scribble to prototype. Certainly, it beholds a temporal trajectory, but the study's focus is on how novelty appears in the specific situations of both objects' becoming. For this focus, it is less important if, for example, one actor built one prototype and the other three. What is important is rather how prototypes assemble previously unrelated elements and how differently actors enact that prototype as new. The depth of data allows the sketching of how actors select and assemble various semiotic and material resources depending on whether a sketch, prototype, working material unit, or research paper embodies an object's novelty. Characterizing the sense of coherence and difference that occurs in those various situations is the matter of this study.

1.3 Theoretical Perspectives: Invention, Differential Pattern, or Biographical Passage

In a general sense, there is a vast amount of literature dealing with the relation of novelty and technology. Some of that literature considers novelty or newness explicitly as a conceptual category, whereas other strands express its occurrence using different terminology. The choice of literature, which I discuss in the following section, is based on two criteria: Firstly, relevant approaches need to capture practices of how novel objects come into being. Secondly, relevant ap-

---

4   See Appendix for a timeline of events and data collection.

proaches should entail lessons learned at an abstract level that allow comparison between science and art. This should allow one to learn something about novelty and technological objects *beyond* production modes that are limited to a particular field. In the following, I will distinguish between three perspectives that meet these criteria but still approach the relation between novelty and technological objects differently.[5] The perspectives are novelty as invention, as differential pattern, and as biographical passage. In order to illustrate differences between these perspectives, I will draw on Leonardo da Vinci's work as an artist and engineer as an example.

1.3.1 Novelty as Invention

Novelty as invention is a perspective that focuses on the achievements of engineering new technical devices. In its colloquial sense, it is much related to the figure of the inventor, whose ingenuity is the source of the technological objects that are regarded as ahead of their time in historical retrospection. Biographical accounts of Leonardo da Vinci's life work are a pivotal example of this. These popular accounts commonly acknowledge his achievements as artist, philosopher, and inventive engineer whose works continuously exceeded the means of its times (i.e. Bortolon 1965; Gibbs-Smith 1978). His paintings are world famous and were regarded as exceptional artworks during his lifetime. Da Vinci was not only original in terms of staging Christian figures like John the Baptist, who he painted smiling and with one finger pointing upwards, but he was also original in his painting techniques. For instance, he developed the chiaroscuro technique, which became quintessential to the dark-light effects of Renaissance paintings (Bortolon 1965). Despite his famous paintings, biographical accounts refer to his creativity as an inventor of technical devices. In contrast to his paintings, da Vinci was only able to realize a few of them, as most preceded the technical means of their time. His sketches document his talent for imagining apparatuses and are not only drawings of fantastic machines but entail detailed explanations

---

5 A different discussion of literature could distinguish, for instance, between disciplines like the sociological versus the anthropological perspective. However, disciplinary provenience does not necessarily entail that perspectives meet the criteria mentioned above. Hence, it is sensible to typify perspectives based on specific criteria as to discuss different approaches concerning a shared problem. Assuming unity among literature of a certain label is also misleading concerning the problem discussed here, because many approaches lie across disciplines like, for instance, Igor Kopytoff's approach that combines anthropological and Marxist notions but still became paradigmatic for studies on biographical passages. Nevertheless, the subsequent text flags out disciplinary origins wherever this clarifies the argument, but does not take disciplinary provenience as starting point nor as result of discussing theories.

1.3 Theoretical Perspectives: Invention, Differential Pattern, or Biographical Passage 33

of their inner workings. Nevertheless, it was both creative talent that pushed da Vinci to plan a wide range of technical applications and the necessity stemming from his assignments as technical engineer in times when Italian principalities spent most of their money on war machinery. In his drawings, da Vinci drafted aeronautic machines that resembled the flight of birds and sketched rotor devices that pre-empted modern helicopter technologies. Not all of his inventions were solitary creations of his mind; for instance, his draft of a parachute picks up ideas from his contemporaries and improves upon detailed aspects of these (White 1968). Whereas some of his technical sketches were actually engineered and tested and led to plenty of injured test people, others were only realized hundreds of years later and proved to be perfectly practicable technologies, like his draft of a diving suite (Gibbs-Smith 1978, 76-7).

The colloquial understanding of the invention as a novel technical device created by an ingenious inventor is somewhat reiterated and somewhat deconstructed in classical texts on the sociology of innovation and related disciplines. These classics usually regard invention as a primary phase in innovation cycles or path models. Whereas diffusion characterizes innovation, invention is a preceding activity that regards ideas, experiments, tests, plans, etc. – all those activities wherein material and organizational environments are unstable. As such, inventions are *not* a central concern of economic studies of innovation. Joseph A. Schumpeter, for instance, who introduced the distinction between invention and innovation, treats invention rather randomly so as to distinguish it from the economically more potent innovation. For him, innovation is the implementation of new practices and technologies, whereas inventions remain ideas or plans and are thus of no importance for economic analysis. This separation was defended by Schumpeter quite ambitiously, and he stressed that, whether inventions emerge autonomously and without any practical need or they respond to a given business situation, still, "the making of the invention and the carrying out of the corresponding innovation are two entirely different things" (Schumpeter 1939, 80-1). However, he acknowledged that the question of whether "necessity is the mother of invention" is difficult to answer, and disciplines other than economics might respond to it differently.

With that statement[6], Schumpeter passed the ball to sociology, which may study the relation of inventive action and technical progress differently. From a sociological perspective, William F. Ogburn was one of the first authors dealing with technical progress, its social causes, and its impacts. Building upon Marx, his famous "cultural lag" hypothesis regards invention as the prime mover of history. That is, technological progress usually advances cultural progress, and

---

6   That statement can be found in a footnote of his (Schumpeter 1939: 81, footnote 25), and Schumpeter further refers to S. Colum Gilfillan.

the social world may only adapt to the faster and farther moving technical inventions that are considered to move autonomously and spontaneously (Ogburn 1964).[7] In this sense, Ogburn regarded inventions as a force of history that stimulates human activity. He defined them as "a combination of existing and known elements of culture, material and/or non-material, or a modification of one to form a new one" (Ogburn 1964, 23). He continued and stated that inventions are the evidence upon which we base our observations of social evolution.[8] In so doing, Ogburn assigned inventions a crucial position in his theory of social change and hence attempted to capture how they come about. For Ogburn, inventions result from three factors: mental ability, demand, and the existence of cultural elements that are combined and formed. Ogburn regarded mental ability as an individual capacity, in the sense of ingenuity, as well as the capability to learn within a social collective. Existing demands do not cause all inventions; they may also be accidental. However, demand directs the learning process, as the use of an invention implies demand. The third factor is related to accumulation, which occurs when activity adds more new elements to the cultural base than are lost (Ogburn 1964, 23-4).

S. Colum Gilfillan continued with Ogburn's account of social evolution and, in particular, advanced the notion of invention. In his book *The Sociology of Invention*, Gilfillan attempted to capture the nature of inventions as well as its fostering factors and effects. In so doing, he continued the evolution analogy. Similarly to Ogburn, he characterized invention as a "perpetual secretion of little details" having "neither beginning, completion nor definable limits" (Gilfillan 1935, 5). This places emphasis on mutation and variation that continues through time in reciprocal dependence on its environment. In this regard, an invention is a complex of diverse elements, such as a design for a physical object, the process of working with it, and the necessary financial and scientific elements, as well as the anticipated infrastructure of its existence (Gilfillan 1935, 5). He exemplified this account in his case study of the invention of the ship. Whereas Ogburn rather carefully deconstructed the genius figure, Gilfillan was more explicit in doing so and stressed that it is a matter of historical writing to attribute someone an extraordinary status. In particular, the Renaissance began to accredit new technologies to single inventors, which increased the social reputation of figures like da

---

7 The cultural lag hypothesis as well as the following account of invention were first released by Ogburn in 1922 in his book *Social Change with Respect to Culture and Original Nature*. I cite a book of selected papers called *On Culture and Social Change*, which was published in 1964.
8 Ogburn is an author continuing and advancing evolutionary theories of social change. In this regard, his terminology, like "observing evidence," is not arbitrary and continues in conceptual accounts of "selection" and "adjustment." In a footnote, he regards "mutation" as the evolutionary pendant to invention in the sense of "a new variation that is inherited." In so doing, he understands invention as the primary force of social evolution.

Vinci (Schmidtchen 1997). Certainly, Gilfillan admitted, there were great men who were the first to add a fourth mast to their sailing vessels or who ambitiously employed the application of steam engines to boats. However, their role is rather a matter of possessing knowledge resources, concentration of attention, and motive for action, and less a matter of exceptional mental capacities (Gilfillan 1935, 71-91). Hence, inventions are new combinations that occur in a stream of variations ("evolution") and less in a series of creations. Later, Gilfillan clarified in response to critics of the cultural lag hypothesis: the sociology of technical progress does not regard technology as having determined consequences in society. Rather, he stresses, the question of whether invention is the prime cause of social change has no answer, since invention as cause is inseparably linked with invention as effect. Causes and effects are inextricably bound up in inventions (Gilfillan 1952, 198).

Historian of technology Thomas P. Hughes reiterated the evolution analogy introduced by Ogburn and Gilfillan, but re-defined the position of the inventor figure. Hughes understood invention as a primary phase within the evolution of large technological systems. In contrast to Schumpeter, Hughes did not understand invention as a subordinate phase, wherein the real action takes place later through the diffusion of an innovation, but acknowledged the strategic foresight under uncertain circumstances that characterizes the inventive action of "system builders" (Hughes 1987). In his concept of evolutionary patterns in the genesis of large technological systems, Hughes distinguished between four activities that each dominate specific evolutionary phases: invention, development, innovation, and technology transfer (Hughes 1987, 56ff.). Invention and development relate to novelty, whereas the latter phases emphasize diffusion. Hughes' story of invention is somewhat a story of heroes, in the sense of men who followed and realized an inventive idea even though their success was far from certain. He distinguishes between conservative and radical inventions. Inventions can be conservative if they improve or expand existing systems, or radical if they do not become components of existing systems but rather inaugurate a new system (Hughes 1987, 57-8). Furthermore, Hughes separated independent inventors from those with professional backgrounds by stressing their choice of problems. Whereas the latter tackle pre-defined problems, the former have the ability to identify problems of larger scope, albeit with less financial support. Alexander Graham Bell, for instance, was an independent inventor who was able to connect his experience as a speech therapist with the technological developments of his time concerning signal transmission. This way, he was able to identify problems of large scope by making a connection within and across specific contexts. During what Hughes regarded as the development phase, the social construction of technology becomes clear (Hughes 1987, 62). Not only are professional inventors embedded in an organization, but independent inventors are also in dialogue with an inventive community. The development phase is when a simple idea that works well as an invention in the mind of its inventor changes into a system that can function in an environment permeated by various factors and forces. It is the time when inventors construct technical experiments and tests, in the sense of artificial environments that successively become more complex and real-world-like (Hughes 1987, 63). Hughes' detailed and analytical historical account sheds light on the contingencies in the development of large technological systems, which were recognized later on as great technical achievements, while equally identifying patterns that continue in different developments. The figure of the inventive entrepreneur, who pushes through his inventive ideas despite all doubt, drives Hughes' account. Novelty, in this regard, is the radical invention in the mind of the inventor. All subsequent phases are a matter of stabilizing the inventive intent.

Similarly to Hughes, Lynn White Jr. sketched the heterogeneity of the causal relationships that signify historical inventions. His historical examples stress the embeddedness of technological creativity and the contingencies of transferring ideas (White 1962). The historical diffusion of technologies may behold delays caused by societal or geographical barriers, or it may be accelerated by certain key events that spread an idea. For instance, the crank is a very simple idea that first appeared in the Han dynasty as a means of winnowing husked rice, whereas an identical apparatus found in the 18th century in Austria revolutionized European mechanics. Lynn's point is that patterns of diffusion tell only a little about a technology's inventive context (White 1962, 492).

These economic and sociological perspectives on novelty as invention entail several remarks, which I would like to sum up in dialogue with the introductory example of Leonardo da Vinci. Clearly, Schumpeter would not have paid much intention to da Vinci's technical drawings. Instead, he would have stressed the innovative effect of da Vinci's drawing technique, which diffused and sustainably altered the method of Renaissance painting. The parachute and helicopter drawings, as well as the prototypical flying machines, would have been labeled by Schumpeter as inventions and, hence, as ideas of no economic efficacy. Ogburn would probably have reacted differently and taken da Vinci's technical drawings as evidence that technical progress is always ahead of its cultural environment. In companionship with Gilfillan, he would have stressed that da Vinci's inventions are continuous alterations of objects that he picked up from his environment and re-combined in a technical draft. From their perspective, his drawings are evidence of a societal evolution rather than of an individual ingenuity. Hughes' work corresponds with the evolutionary perspective, but still acknowledges da Vinci as an extraordinarily radical inventor. In Hughes' view, da Vinci's technical drawings behold radical inventions, as his drafts did not become part of existing systems, but would go on to inaugurate new systems, even if they were only realized hundreds of years later. In this respect, Lynn could prove his point and explain that, retrospectively, people attribute the invention of the parachute to da Vinci, but the invention relates to the Renaissance's fascination with flying.

In this respect, these studies hold important lessons in terms of the historical conditions and effects of inventions. However, they have limited scope and may not account for practices that enact objects as novelty. This perspective's only means is to retrospectively account for an invention's trajectory. Situational practices, like material or communicative interactions concerned with constructing something coherent, exceed the perspective's scope, because it limits an object's existence to its technical documentation. In terms of my example, the cited scholars leave open the questions of what role the playful abstractness of drawings played in da Vinci's epistemic practices, how he interacted with nature so as to translate particular characteristics of birds into flying apparatuses for humans, and how he used his body or other situational resources at hand to present his work as extraordinary invention. However, these diverse elements are crucial to delineating how novelty becomes part of a shared reality despite an object's technical documentation.

1.3.2 Novelty as Differential Pattern

The second perspective treats novelty as a differential pattern occurring in epistemic practice. In particular, it signifies the social theories of American pragmatism as well as science and technology studies. Both strands share a focus on interactions and interdependencies between the material resistances of the world and the image humans construct of the world. I will return to my example of Leonardo da Vinci to illustrate differences from novelty as invention.

According to American pragmatism, practical consequences, efficacy, and the bodily experience of interacting with the world constitute meaning-making. This perspective, shared by scholars from the Chicago School as well as their forerunners Charles S. Peirce and William James, not only signifies the constitution of the human self and society (Mead [1934] 1967), but also the experiential inquiry into the world in science and art (Dewey 1938; Dewey [1934] 1980). John Dewey understood inquiry as "the controlled or directed transformation of an indeterminate situation into one that is so determinate in its constituent distinctions and relations as to convert the elements of the original situation into a unified whole" (Dewey 1938, 104-5). In this sense, disturbing, troubling, ambiguous, confusing, conflicting, or obscuring situations are transformed into determinate and unified situations through an interactional response that takes all previous situations and their similarities into account (Dewey 1938, 105ff.). With this understanding, Dewey focused on the interactional transformation of loose elements into problematic objects, which makes them the matter of investigation. For Dewey, inquiry goes beyond scientific procedures, but regards the social nature of apprehending the world.

At a societal level, Helga Nowotny described this apprehending of the world as driven by an "insatiable curiosity" that drives the continuous quest for novelty. This striving for newness is not limited to a specific field but lies across art, economy, and science as a cultural phenomenon fueled by the ability to imagine the future. Nowotny regards our curiosity as insatiable, "first, because the space of possibilities and reality that is to be explored still approaches infinity; and second, because more and more means and instruments [...] are at our disposal to expand the space of our experience" (Nowotny 2008, 3). In her account, the new appears in two variants: firstly, as a recombination of pre-existing and known elements or, secondly, as a break that underscores the contrasts with that which already exists (Nowotny 2008, 11). Nowotny does not explicitly draw on pragmatist theory, but she has illustrated how Dewey's thinking can be comprehended at a societal level, too – as an insatiable fascination with endeavoring towards the new.

On a micro level, Werner Rammert picked up Dewey's inquiry pattern and advanced it toward a learning concept of "experimental interactivity." He regards experimental interactivity as a learning process, characterized, on the one hand, by increasing control over objects and, on the other hand, by enhancing meaning that is culturally situated. According to Rammert, through the circularity of interaction, events are translated into objects and meaning is produced as a category of material behavior (Rammert 1999). Meaning is a translation of differential patterns into objects in light of culturally and biographically framed situations. Coming from a gender- and Foucault-inspired discussion, Karen Barad reiterated this concern of linking matter and meaning in her agential realism. Barad has urged that there is not matter without meaning, as both always co-produce through "intra_actions." In her account, meaning is not an ex-post construction, but is already inscribed in experimental settings whose technologies produce boundaries that render novel entities (Barad 2007, 175).

"The Mangle of Practice" is a concept proposed by Andrew Pickering, who has referred to its pragmatist foundations explicitly and, in this vein, focuses on the circularity of resistant events. Pickering has proposed a performative understanding of science instead of regarding scientific knowledge as representations of nature. Hence, he wants to capture how science produces knowledge in real-time and not what counts as scientific discovery retrospectively. His pivotal concern has been material agency and how its resistance affects the production of new knowledge, which Pickering does not regard as determined *a priori* but as emerging through interaction. In a "dance of agency," human and non-human entities approach each other and enter a dialectic of resistance and accommodation. Whereas resistance is the failure to capture agency in practice, accommodations are active human strategies that respond to that resistance. These strategies can be rather different: they may seek to integrate the experienced resistance into concepts, they may lead to changing goals and intentions, or they may motivate a change in the material setting of experiments and tune technical apparatuses and instruments (Pickering 1995, 22). In Pickering's understanding of scientific practice, experimental results are not passive discoveries, but are produced through continuous chains of interactions that trigger differential patterns. In so doing, he not only refers to the knowledge produced as novelty, but also the form it enters.

The pragmatist understanding of novelty sheds a different light on my example of da Vinci's epistemic practice. Instead of regarding him as an ingenious inventor as in biographical accounts (which I have already deconstructed with the help of evolutionary approaches to invention), now, through pragmatist glasses, his experimental practice comes into focus. Wolfgang Krohn gave an account of the Renaissance and "novel" science in this regard. He stressed that da Vinci understood the experimental method as the only method that may proliferate approved empirical knowledge. This is shown in the notes that accompanied his technical drawings, which give detailed accounts of experimental setups, like those for testing the load capacity of artificial wings. These documents show how knowledge of nature and technical knowledge are interrelated in da Vinci's practice and, to Krohn, how the controlled construction and observation of events became his primary epistemic method (Krohn 1977, 77-8).

Experimental setups are also in the focus of the empirically orientated laboratory studies, which similarly aim at deconstructing scientific knowledge as a representation of nature. Laboratory studies claim that scientific objects do not already exist as entities in nature that are then passively discovered by scientists, but instead are actively constructed in the alliances between scientists, epistemic cultures, instruments, infrastructure, discourses, and organisms under investigation (cf. Callon 1986; Latour and Woolgar 1986; Latour 1987; Knorr-Cetina 1980; Knorr-Cetina 2000; Rheinberger 1992; Collins 1982; Pickering 1995; Barad 2007). Bruno Latour and Steve Woolgar emphasized in this regard that they do not conceive of scientists' activities "as pulling back the curtain on pregiven, but hitherto concealed, truths" (Latour and Woolgar 1986, 129). Rather, the dirty work of scientific practice constructs objects; it assembles and re-arranges elements to produce the entity under investigation. According to their study on the peptide TRF (H), Latour and Woolgar claim that the difference between expected and unexpected signals constructs new objects. Differences such as those between two graphs become an object through the accumulation of technology that stabilizes previously random patterns. If scientific technologies identify distinction and reproduce it, they stabilize substances, organisms, particles, etc. as stable, distinct, and new. Infrastructures, discourses, and scientific labor are all inscribed in the differential patterns that they produce and that they translate into novel objects. In this respect, laboratory studies deconstruct the border between the world and discourse, nature and technology, and subjects and objects by stressing continuous, step-by-step translations that constitute the existence of entities *as* objects.

The enactment of boundaries between technical instruments and objects under investigation is also the focus of Hans-Jörg Rheinberger's experimental systems. For Rheinberger, experimental systems are the basic unit from which experimental reasoning proceeds. They constitute scientific objects within specific technical conditions. Experimental systems must be capable of differential reproduction, which means the repetitive production of things that go beyond our present knowledge – "that is to behave as a 'generator of surprises'" (Rheinberger 1992, 307). In that sense, productivity means, for experimental systems, generating reproducible differences that are not so much answers, but foremost materialized questions. Rheinberger distinguishes between two inseparable structures or components within experimental systems: epistemic and technological objects. Epistemic objects are the entities under investigation. They cannot be fixed from the beginning and present themselves in a irreducible vagueness, as "one does not exactly know what one is looking for" (Rheinberger 1992, 310). In contrast, technological objects are the stable elements in experimental systems. They make up the infrastructure that needs to work in a determined and predictable manner. The crucial note in Rheinberger's concept is the relation between epistemic and technological objects, which is not a material one, but a functional one. Hence, both object types do not describe fixed or well-defined parts of a system, but change their boundaries depending on their enactment within knowledge activities. This includes the tuning of technical instruments as one part of scientific work that is not acknowledged through scientific reputation but, nevertheless, requires sophisticated knowledge activities, as Harry Collins outlines in his study on experimental physics (Collins 1982). Novelty, in this regard, is the outcome of experimental systems – the differential pattern produced through technical structures and translated into graphs, equations, or other representations.

Approaches that are more recent show that not only are differential patterns technically produced, but bodily interactions also co-produce epistemic objects. In these studies, bodies are more than a means to enact human agency. Instead, they become epistemic instruments through which a scientist engages with an object under investigation. In this regard, Morana Alač has spoken of "bodies-in-interaction" as the interactional articulation of bodies in situated epistemic practice. She has observed, within the practice of robot researchers, that it is not only mimicry when scientists act as if they were a humanoid robot; the scientist's body materially informs and guides programming and is employed as instrument and model for engineering the robot. In practice, the practitioner's body aims at "becoming a general type of the human body" as it is displayed to be collectively observed in the shared environment of the robot laboratory (Alac 2009, 520-1). This bodily simulation is a distinctive form of enacting epistemic objects that have not yet materialized but are referenced in communicative situations. Natasha Myers has called such body-work "embodied imagination" (Myers 2008, 165). In her study on protein modeling, she argued that material and mental models are not dualistic, but deeply entwined. Through embodied imagination, researchers incorporate the inner structures of models and enact these as epistemic objects. They use their bodies to make graphical objects tangible, and they employ gestures and movements in communication with novices in order to flesh out and relay their knowledge about otherwise only virtual objects (Myers 2008, 180). These studies show how novel objects are situatively rendered through such diverse realms as technological artifacts, symbols, and bodies.

Approaches used by laboratory studies raise certain skepticism concerning my example of da Vinci and the historical sources one has to rely on. Their real-time ethnographies deconstruct the superior position of the scientist and show how heterogeneous elements connect at the micro-level and how humans and non-human agents take part in constructing novel objects. Figures like the inventor and approaches that single out human capacities for proliferating the new do not seem appropriate in this light, as they do not acknowledge the contingency of practice and the networks of human and non-human alliances that form new objects. Nevertheless, laboratory studies also urge us to look at places, and, in so doing, they bring the Renaissance workshop into focus. The workshops of the Renaissance were run by artists and craftsmen and, hence, were places of collective material labor. Although there is not much knowledge about da Vinci's workshop, as he worked in many different places, he still needed very distinct facilities for his practice – for example, facilities for conducting studies on the anatomy of the human body. He dissected many human bodies and drew very detailed studies of muscles, organs, and even the first drawing of an embryo in the mother's womb. Certainly, there is no way to find out the details of his practice, but, from the laboratory studies perspective, the practical labor of dissecting a body and its translation into an anatomical drawing becomes significant, because the rendering translates the body into an object that can be related to other objects. As such, the drawing creates an image of the general human body, not just one of the specific dead bodies on da Vinci's table. Furthermore, the drawing may be related to other anatomical drawings that came before. In this respect, difference is enacted through relating renderings of the world that are co-constructed in alliances between humans, discursive technologies like drawings, and the material world. In Latour's words, and from a laboratory studies' perspective, an inquiry into novelty would follow the drawing and not the biography of a human inventor.

To sum up, American pragmatism and laboratory studies argue in very similar ways concerning the relation between novelty and technological objects.

They focus on immediate interactions and are particularly concerned with their material and bodily dimensions. This does not mean they put meaning aside. Rather, both consider meaning as a matter of accommodating material behavior and translating experimental events into objects. This crucially implies changing object statuses. An object is not a fixed entity, but requires continuous effort so as to be stabilized as a new object. The role of technology is fluid in this respect. Epistemic and technological objects may change their statuses depending on their functional enactment in experimental settings. Hence, technologies are not mere means to render a new object visible, but are always inscribed in what the scientific object actually is. Separating an object from its constitutive infrastructure is then an agential cut that makes entities become visible as objects. Such cuts may lead to translating differential patterns into subsequent objects, like graphs, curves, technical renderings, etc., or to disciplined and law-like functionalities, such as a new technical capacity. In this regard, technology relates to novelty as material resistance and the infrastructural trigger of differential patterns, as well as a substrate that renders differential patterns visible.

However, the perspective is limited by its narrow focus on experimental practice. In particular, how long-term semiotic structures interweave with material practice is not in the scope of laboratory studies. In my da Vinci example, this is, for instance, how previous stylistic eras, like Gothic art, structure the specific difference of style in da Vinci's Renaissance paintings, or how da Vinci's portraits of saints reiterate as well as slightly modify traditional Christian figures. All this conceals the focus on the patterns that emerge through small-scale interactions. Despite considering discourses as inscribed in technical settings, there is no conceptual space for the inclusion of stories, imaginaries, figures, symbols, or visual codes in the enactment of novelty.

1.3.3 Novelty as Biographical Passage

A different way to address the relation between novelty and technological objects is offered by biographical passages. I use "biographical passage" as a label to summarize approaches that investigate the long-term processes of the collective and individual meaning-making of objects. In Arjun Appadurai's words, it is about the "social life of things" and their journeys through different contextualization and valorization (Appadurai 1986). The biography metaphor is used widely in this respect (Kopytoff 1986; Daston 2000; Dant 2001; Hoskins 2006). In anthropology, there is a certain tradition for the biographical studies of things, stemming from the influential work of Marcel Mauss and his cultural studies of gift exchange (Mauss 1966).

1.3 Theoretical Perspectives: Invention, Differential Pattern, or Biographical Passage 43

Coming from this tradition, Igor Kopytoff spoke of the "cultural biography of things" in order to address how objects receive and change their values through market exchange (Kopytoff 1986). He conceptualized two basic mechanisms: commoditization and singularization. Commoditization is to attribute things a specific exchange value in order to enter market relations. In this process, things are decoupled from the labor of their production and enter new relations with other things, whose value is determined through market relations (cf. Marx 1976). This process creates homogeneity among things with previously complex natural origins. Singularization, in contrast, makes things special, unique, and shine out of the standardized sphere of commodity exchange. For Kopytoff, culture ensures that some things remain unambiguously singular and resist commoditization, or, respectively, are re-singularized through changing cultural contexts (Kopytoff 1986, 73). Singularization of the latter kind is valorization that is not based on fiscal value, but on cultural forms of symbolic value. This is when social values, such as moral standards, cultural heritage, or originality, are ascribed to the object. In certain societies, manioc is one such example, as it has the material form and availability of a regular market commodity, but is kept from fiscal exchange due to its sacral value. From Kopytoff's perspective, novelty is a matter of singularizing an object. It is not so much resisting commoditization, which Kopytoff regarded as a matter of power, but more the ascription of a unique value that separates a new object from a profane object.[9] For such attribution, novelty must be somewhat represented in the value sphere of a specific societal field as originality in science (cf. Polanyi 1962) or aesthetic affect in art (cf. Reckwitz 2012).

Lorraine Daston picked up the biography metaphor, too, and used it to capture how scientific objects come into being (Daston 2000). She re-ordered the essays of her edited volume regarding four modes of the becoming of scientific objects: salience, emergence, productivity, and embeddedness. Although Daston has used the biography metaphor explicitly, she has not conceptually related the four modes to a sociological understanding of biographies. Nevertheless, salience and embeddedness make a point in terms of biographical novelty, as they refer to collective meaning-making. Daston has written that salience is not so much about absolute novelty, but more about the historical increase in attention

---

9   Tim Dant has discussed Kopytoff's use of the biography method and has made a reflexive argument concerning the engagement of the social sciences with objects. He has argued that objects do not have to be identified as singularized in terms of their exchange or ritual value, but the task involves taking a particular object and making it singular through the process of writing (Dant 2001: 11). In contrast to Kopytoff, Dant singularizes profane objects and writes biographies about their embeddedness in a specific social world. Nevertheless, Dant does not entail remarks on an object's cultural novelty; rather, they inform methodological appropriations of object biographies.

paid to particular objects. This implies the multifarious ways in which previously profane objects are transformed in their meaning and become objects of scientific inquiry (Daston 2000, 6). Essays in her volume stress how colloquial objects, like dreams, monsters, or identity, are re-interpreted through scientific interest. This process is foremost a transformation of meaning that classifies colloquial phenomena into scientific categories. Through a change in meaning, objects become coherent categories of investigation. However, that does not imply a concealment of semiotic relations; rather, specific cultural circumstances charge strange facts with significance and forge connections between them. The capacity of objects to create new relations is also emphasized by their coming into being through embeddedness. The embeddedness of a new object's becoming stresses the potential of science to create new objects, as well as how new objects enforce "new techniques, differentiations and associations, representations, empirical and conceptual revelations" (Daston 2000, 13). The more widely scientific objects connect to other phenomena, the more they yield layers of hidden structure. Daston wrote that embeddedness "captures the distinctively generative, processual sense of the reality of scientific objects, as opposed to the quotidian objects that simply are" (Daston 2000, 13). Her account is strict in stressing the historicity of scientific objects and describes the changes in meaning that go along with it. Nevertheless, her object biographies remain unconceptualized in their analogies to human socialization.

Sherry Turkle is somewhat more rigorous regarding the analogy of biography and object-meaning. For her edited book *Evocative Objects*, she collected examples of intimate relationships between people and things. These things are very different in terms of their material compositions, or, if they are scientific or quotidian things, what they share is an intimate relationship to biographical passages in personal human lives. The stories in Turkle's book report on biographical passages in which certain, dear things gain specific meaning. Turkle concludes from these stories that it is particularly the "things we think with" that evoke intellectual and emotional meaning. This evocation of thought relates to a person's emotional life and biographical memories attached to an object, or it relates to a thing's resistance to fitting into established categories, which evokes intellectual engagement. The latter, for instance, is reported in a story about Chinese scholars' rocks. These rocks evoke scrutiny of their origins, as it is not easy to tell from their appearance whether they are handcrafted or natural objects. By placing them on a pedestal, they become cultural objects that foster philosophical engagement with their peculiar existence. Reframing them in a biographical context and in relation to Western philosophical traditions causes them to challenge existing categories such as nature/culture, east/west, or science/identity (Rosenblum 2007, 252ff.). Turkle wrote that it is often things on the boundaries that are disruptive and are sources for new ideas (Turkle 2007, 322). Her perspective on intellectual engagement with objects is active, as she stresses the role of tinkering, play, and material practice in changing an object's meaning. Donald W. Winnicott stressed, in this regard, how children build a self by playing with toys, which makes them relate to their surrounding world (Winnicott 2005). In the essays in Turkle's book, this also holds true for adult life, as shown in one story in which the author reports on how crafts fostered her understanding of math. In these examples, material and abstract practice mesh through interaction – in objects, "the abstract becomes concrete, closer to lived experience" (Turkle 2007, 307). Novelty, in this regard, is a matter of re-framing objects through active engagement and drawing connections between previously separated realms. These meaning-making processes are individual because they involve personal biographies, just as they are collective because they reflect cultural contexts.

The notions of biographical passages from Kopytoff, Daston, and Turkle share a certain favoring of re-contextualizing objects so as to make them singular, scientific, or evocative. They stress the potential for individuating objects through creating new connections and cultural framings. There is a nice example from da Vinci's mechanical drawings illustrating this point. One drawing of his, supposedly, pre-empts the invention of the helicopter. It shows a whirling sail connected to a mechanical rotor. Nevertheless, it would be wrong to call da Vinci the inventor of the helicopter, as he adopted the design of a much older Chinese toy for his drawings, and, furthermore, his apparatus only indirectly inspired the aeronautic pioneer Igor Sikorsky to develop technical designs of helicopters (cf. White 1962; Gibbs-Smith 1978). This kind of meaning-making through new connections across large timescales have been put into focus through the notions above and have not come to the forefront via the perspectives of novelty as invention and differential pattern.

However, such re-contextualization entails a backlash in terms of an object's valorization: in his book *The Shape of Time*, George Kubler considers the historical sequences of art works and the continuity of change across time. He states that considering artistic intent and style is somewhat limited for understanding the cultural value of art. Instead, he proposes placing art objects in larger continuums and stressing their duration throughout history. Art objects behold specific positional values that stem from their systematic positions with respect to a pertinent sequence (Kubler 2008, 89-90). In this sense, every object not only has a position in time but also a position in a system of form, to which its occurrence belongs. This refers to, for instance, the position of Zurbarán's portraits of the twelve Apostles; each of them are standalone portraits, but they also form a coherent work, as their gestures are signified through their position as a group. In this regard, Kubler shows how history matters for the meaning of objects, but recontextualizing objects does not necessarily entail valorization, but instead may also lead to a loss of meaning.

With this critique in mind, I want to include another notion of novelty as biographical passage, proposed by media philosopher Boris Groys. In comparison, Groys is most explicit about novelty and takes novelty as a theoretical concept. He does not apply the biography metaphor, but instead takes "the archive" as *the* significant passage point that marks an object as novel. On the one hand, the archive stabilizes novelty, as it preserves the qualitative difference of an object from all the other objects that came before and that are already included in the archive; on the other hand, in so doing, the archive drives the quest for the continuous proliferation of novelty, as it prevents achievements from losing their meaning through time. Groys has written that only when novelty does not pose a danger to identity and tradition may it become a positive and collective demand (Groys 1992, 23). The archive is the technical and medial infrastructure for preserving the proliferation of difference and makes inventive achievements commonly accessible. Hence, entering the archive is a matter of valorizing an object as novelty, since it is marked as a relevant contribution. This implies an imperative for creating novelty, or, as Groys wrote: "The first law of the archive is: you shall not do the old!" (Groys 1997, 31, own translation). In this sense, the archive says what you should not do, which is repeating what is already included, but it does not determine what objects or what forms are valorized as novelty. Groys' pivotal example is the museum, which is an institution that collects art works and which marks them through their incorporation as relevant contributions to art. This includes objects becoming a matter of discussion through archiving, as they are related to an archive's stock. In this respect, archives are symbolic technologies because they embody cultural meaning and create modes of comparative conservation (Nowotny 2008, 28). Nevertheless, in Groys' account, it remains unclear how the expanded set of modern symbol-technological instruments of preservation influence modes of archival valorization.

At first sight, Groys' archive does not seem to behold too many insights for my da Vinci example. Clearly, da Vinci's works are part of the societal memory and are valorized as novelty. The institutions of art history have adjudged them distinctively different from what came before in terms of artistic style as well as technical engineering. Still, Groys' account beholds a certain point: valorization is less about forms and more about difference. Whereas da Vinci's paintings had already earned him a strong reputation during his lifetime, his drawings were accounted as valuable much later, after they were related to previous and following technologies. Through the historical juxtaposition of his drawings with other sketches, patents, and plans that entered the archives much later, the progressiveness of da Vinci's drawings has been accounted for. It is not so much the form that is crucial to such valorization, but assessing that the drawings do not resemble the old. This point, which Groys makes, makes a clear distinction from novelty understood as a differential pattern that regards forms as essential to how differences are constructed.

To sum up, novelty as biographical passage stresses the relation of collective and individual meaning-making. Certainly, most approaches summarized here do not take this metaphor seriously, as they hardly entail concepts analogous to human socialization (cf. Dant 2001). Nevertheless, the approaches share certain points. Object biographies signify the tension between, on the one hand, becoming part of a collective of objects that fosters comparison and exchange, and, on the other hand, increasing an object's value through singularization. This tension between commoditization and singularization signifies market relations as well as emotional relationships between humans and things; at both levels, novelty occurs as new meaning through singularization. Time is a central dimension in object biographies. Over the course of time, objects may become a matter of changing interests, which conflates different cultural realms or connects them in a new way. This happens, for example, when colloquial or mystical objects become a matter of scientific interest, or, inversely, scientific objects become a matter of artistic interpretation that leads to novel art objects. Nevertheless, meaning stabilizes through biographical passages when they valorize an object as a relevant contribution to societal memory. In relating an object to previous objects, their difference qualifies an object for entering the archive. Entering an archive stabilizes an object's value as novelty while simultaneously allowing further meaning-making through discussion of how it relates to the archive's stock. In this sense, biographies signify passage points that mark an object as novelty in relation to a collective of objects.

Nevertheless, object biographies are also limited. For instance, referencing and meaning-making are very selective, and only certain aspects of an object become the matter of new interpretations; however, biographical passages tell only a little about the *processes and practices* of selecting. This includes how authors actively translate or change the material composition of an object to fit symbolic technologies of valorization. In this respect, biographies need to entail activities of writing biographies and their material labor instead of only retrospectively sketching a trajectory of changing meaning. This issue can be approached by taking the biography metaphor more seriously and stepping into dialogue with related concepts. For instance, George H. Mead emphasized that bodily interaction constitutes the human self. The following question is then, what can we learn from the analogy between developing a self and object biographies?

1.4 Toward Novelty as Articulation

The preceding discussion aims at sketching differences concerning perspectives on novelty and technological objects. All three perspectives focus and conceal different aspects and analyze technologies in diverse settings and on different time scales. Hence, they all contain specific shortcomings as well as lessons learned. Two central points summarize the perspectives' shortcomings: Firstly, the perspectives insufficiently acknowledge the diversity of realms through which an object becomes novelty. Each perspective's focus comes with a blind spot, and none of them considers in one heuristic the rendering of novelty through material tinkering or through collective meaning-making based on signs, values, imaginaries, and the interweaving of those diverse realms. Secondly, if a perspective concentrates too much on experimental practice, it neglects structural influences, and, when a perspective concentrates too much on trajectories, it leaves out the dirty work of experimenting, tinkering, and the entailed contingencies. None of the perspectives captures practice *and* structure or their interrelations in the becoming of novel objects. This is, for instance, how technical infrastructures become resources or impediments in experimental practice, or how situated enactments of single objects reiterate societal imaginaries about technologies.

Nevertheless, the perspectives share a central issue. This shared problem regards novelty and its relation to technological object: *how does an object become an individuated unit and simultaneously become related to what came before?* This is the crucial tension of novelty and runs implicitly through all of the accounts discussed. It is the tension between individuating and relating technological objects, between becoming one and becoming one of many. This fundamental problem is addressed by evolutionary approaches as a tension between continuity and mutation, by differential patterns as a matter of triggering and translating events into new forms, and by biographical passages as singularizing commodities. In this regard, novelty is more than merely the difference from what existed before. Instead, novelty is tied to individuation, to making an object effective and accountable as a unit. Perceiving something as different, as a distinguishable unit, requires separation from the surrounding profanity, whether it be on a material or semiotic basis. Novelty is a matter of coping with the tension between individuating and relating an object – between making an object work and making it meaningful in a distinguishable way, and opening it up for new information with which to connect. This problem is the main issue of this study.

Bringing this tension into the focus of this study not only stems from reflecting theoretical perspectives; the problem is also enforced by initial empirical observations. My first ethnographies and conversations in the RBO Laboratory were only marginally concerned with trajectories of robotic hands as such. How *differ-*

ent the RBO Hand was from other robotic hands was of secondary concern for Deimel and the institute's director, Brock. What mattered primarily to them was how to exploit the material capacities of silicone for grasping. Their activities and their accounts of their research practice were mostly concerned with designing and engineering a hand that worked – as simple as that. Similarly, Baecker only marginally referred to other media artworks when he spoke about ideas for his new installation, which later became *Mirage*. In addition, his activities in the studio focused on realizing technical and mechanical circuits, instead of caring about whether a certain technique opposed common procedures of artistic work. In this regard, my early observations stress that, in the first place, Deimel's and Baecker's activities are less concerned with differentiating a new object in opposition to what came before; on the contrary, both individuals' work is primarily concerned with a more basic problem: creating a material unit that works; *making materialities do something that they have not done before*. Empirically, this is means creating a bending form that allows exploitation of the capacities of silicone so as to grasp items, or creating a functional link between the Earth's magnetic field and the deformations of a mirror foil. These issues are aimed at individuating an assemblage of previously separated things towards a functioning unit.

In this respect, the focus on the tension between individuating and relating and making it an explicit concern for studying novelty drives my proposal for a different perspective on novelty and technological objects. Taking lessons learned and problems into account, the main question to guide a new approach is *how to analyze the ways in which objects individuate and relate through diverse realms considering situational practices and long-term structures*. This question focuses on activities of individuating and relating, while still being concerned with structuring conditions.

In the following, I sketch an approach to this question. The central notion of that approach is articulation. After sketching the basic meaning of articulation and how two selected theoretical strands advance the notion, I outline the elements that articulate technological objects. These elements are figures, technicity, and enactments. Whereas figures and technicity relate to the structures of processes, enactment enfolds contingency at the practice level. The theories from which I build my approach are not fully correlated – neither in the sense that they are based on a common social theory, which would have made the terms they use fully consistent, nor in the sense that their argumentation is directed toward a shared objective, as in, for example, explaining how novelty emerges. What connect the theories are their offerings to understanding the paradox of relating and individuating. The theories that inform my approach stress, on the one hand, how objects individuate and become accountable as a unit and, on the other hand, how relations to specific semiotic and material environments determine an object's existence. Both are entangled processes through which objects become significantly distinct. They are abstract processes, underlying what I consider articulations of novelty.

1.4.1 Articulation

Proposing articulation as a central notion is inspired by the colloquial understanding of the term. In English, the word has a twofold meaning. In the first instance, it refers to the action of distinctly expressing an idea or feelings with words. Here, the word refers to the clarity of sound, the sharp edges of speech that make words distinguishable units. However, in a second instance, articulation also refers to the connection of two parts by a joint. According to this definition, an articulation is the joint or juncture between bones and cartilage in the skeleton of a vertebrate. It is the movable joint between two rigid parts. In this twofold meaning, articulation addresses the tension between individuating and relation, which I regard as the core tension of novelty. It captures ways in which something is expressed as a distinguishable unit as well as how something is related to other units that may form a continuous sequence of joint elements.

However, articulation as a term is somewhat occupied in social science. In order to elaborate on what articulation implies for the study of novelty, I have sketched two directions that advance the term as a theoretical concept. I begin with its understanding in cultural studies and then sketch its understanding in science studies.

In cultural studies, articulation is associated with the work of Stuart Hall, who built upon Antonio Gramsci and Ernesto Laclau and continued developing their Marxist notion of articulation. Hall began explanations about the meaning of articulation by referring to the same double meaning of articulation that caused me to use this term. He explained that, in English, articulation not only refers to distinctive lingual expressions, but also to material connections between two entities, like between a truck and a trailer, which *can* be, but is not necessarily connected as an articulated lorry (Hall 1986, 53). Hall's main conceptual reference is Laclau, who advanced articulations as a theoretical perspective to escape the traps of reductionism in Marxist theory. Criticizing reductionism stresses Marxists' reduced explanations of social structures as determined by class or production mode (cf. Slack 1996). Laclau took this accuse seriously and called for a more theoretical foundation for Marxist categories. Stressing articulations is then a matter of addressing the process of abstraction between the theoretical accounts of discourses and the categories that inhabit such discourses. Class, for instance, is a theoretical delineation of a particular societal formation that stands

1.4 Toward Novelty as Articulation 51

aside from other possible determining forces, instead of a concept within political practice (Laclau 1977, 10). In terms of novelty, this means that novelty is not necessarily an actor's label used within robotics or art discourses. Rather, novelty is a theoretical delineation of a specific formation that connects diverse elements. It is an account of situations from a specific perspective.

In Hall's work, articulation is advanced as a conceptual term to characterize how class, race, political movements, ideology etc. are not determinate consequences of societal formations, but are articulations of diverse elements that could also be otherwise (Hall 1986, 53):

> "An articulation is thus the form of the connection that can make a unity of two different elements, under certain conditions. It is a linkage, which is not necessary, determined, absolute and essential for all time. [...] So the so-called 'unity' of a discourse is really the articulation of different, distinct elements that can be re-articulated in different ways because they have no necessary 'belongingness.' The 'unity' which matters is a linkage between that articulated discourse and the social forces with which it can, under certain historical conditions, but need not necessarily, be connected."

An articulation, in this regard, is the form of connections that can create a unity of two different elements under certain conditions. Hall puts unity in brackets so as to emphasize the contingency of articulating elements. Connections can always be different, and parts may fit together in unforeseen and complex ways. However, when they come together, they create a sense of coherence, a sense of unity that makes their connection meaningful. For instance, religion does not necessarily connect to state power, but both can articulate a non-secular ideology. The articulated unity is a linkage between discourses and social forces, which can, but need not necessarily, be connected. Hall stressed that the unity formed by articulations is always a complex structure, in the sense that things relate as much through their differences as through their similarities. Jennifer Daryl Slack and J. Macgregor Wise clarified regarding Hall that "articulation can be understood as the contingent connection of different elements that, when connected in a particular way, form a specific unity" (Slack and Wise 2005, 127).

In this heuristic sense, Hall urges that studying articulations needs to delineate mechanisms that connect dissimilar features, because no homology is given. This entails searching for connections that are made: how practices are linked to effects, texts to certain meanings, meaning to particular aspects of reality, experience to politics, etc. (Slack 1996). These links are drawn across diverse realms, such as symbolic structures, institutions, and embodied ways of habitualized action. They are not free-floating connections among autonomous entities, but are themselves articulated within larger structures. Articulations are not random associations, but structured combinations. In this regard, delineating them must entail the structured relations of dominance and subordinations in which they occur (Hall 1980, 325). In terms of technology, this implies heterogeneous elements at a practice as well as a structural level. Articulations go beyond physical arrangements that carry the label "technology" and imply contingently related activities, experiences, and affects at a practical level as well as the structures that constitute a technology's existence, like material infrastructures, discourses of power, shared symbols, and institutions. Articulation is the point of convergence when these elements come together and create a sense of unity among themselves.

Hall's understanding of articulation, as it is briefly sketched here, somewhat entails novelty implicitly. It highlights the temporality and contingency of social formations and, furthermore, the appearance of units less as a matter of shared characteristics and foremost as a matter of relating diverse elements so as to enact *a sense of* unity. This understanding of articulations focuses on exactly the same tension between individuating and relating that I regard as signifying novelty.

In science studies, articulation is similarly regarded as a matter of combining elements so as to make them appear unified. However, scholars advance the term in a different conceptual vein than in cultural studies. Bruno Latour and Donna J. Haraway, who both share a general interest in the making of alliances, pick up the term. They share this interest with cultural studies' appropriation of the term. However, their perspectives focus on particular connections between human and nonhuman actors that are articulated (as) discrete units. Latour included articulations in the glossary of his book *Pandora's Hope*, but without giving a substantial definition of the term (Latour 1999, 303):

> "Like translation, this term occupies the position left empty by the dichotomy between the object and the subject or the external world and the mind. Articulation is not a property of human speech but an ontological property of the universe. The question is no longer whether or not statements refer to state of affairs, but only whether or not propositions are well articulated."

Whereas this glossary statement in itself does not clarify what processes he wants to capture, articulations can be understood better by following how Latour has advanced the term throughout his studies. One of Latour's general ambitions has been characterizing scientific practice, not as a passive finding of truthful facts through the scientist, but as an active production of accounts *as* facts. Articulation is then a term that captures how scientists make and craft alliances with various elements, like non-human laboratory apparatuses and entities under investigation. From his study about Pasteur's "discovery" of the lactic acid ferment, Latour advanced articulations as a conceptual term (Latour 1999, 133ff.).

1.4 Toward Novelty as Articulation

In his exemplifying case, his main issue is that the lactic acid ferment becomes a presumably independent entity the more work Pasteur puts into rendering it visible. In this sense, the ferment is not simply captured with a lingual account of its given characteristics, but is produced as an entity through connecting probes, instruments, and other laboratory facilities, as well as agriculture and national politics. All these entities are connected step-by-step so as to produce the ferment as an entity with increasingly significant attributes. In order to conceptualize this process, Latour has proposed a terminological duo consisting of propositions and articulations. Propositions are actants in the sense of occasions or opportunities to bring different entities into contact. Articulations are then connections between propositions. These connections are not statements that capture the world through language with a huge gap between both. Rather, Latour proposes articulations as connections made between several different entities and within heterogeneous material realms. Through connecting propositions, actants become visible as distinct entities. This is not a process like capturing the world with words, but a process that fabricates the world step-by-step so as to align matter with the forms of discourses. In this sense, Pasteur actively articulated the ferment in his laboratory as he connected chemicals and instruments, as well as lingual propositions such as regarding the ferment as a living entity. Articulations connect propositions as differences to "make new phenomena visible in the cracks that distinguish them" (Latour 1999, 143).

Haraway makes use of the term "articulations," too. However, she does not advance articulations conceptually as Latour does. Nevertheless, her use of the term is similar to his. Her realm is in the discourses of technoscience and the ways figures and tropes repeat in the material and bodily corporealizations of modern science. Articulation, in this sense, is a view in opposition to representing science, is a more active term that addresses "clusters of processes, subjects, objects, meanings, and commitments" that are fabricated through scientific discourses (Haraway 1997, 63). In her book *Primate Visions*, for instance, she rendered how the advent of primatology was an articulation of scientific normativity, colonial Orientalism, and masculinity. These articulations manifest in iconic images that, for example, figure apes as humanlike animals, as well as in the material practices of laboratories that reproduce apes as a species akin to the human (Haraway 1989). As with most of her terminology, "articulations" entail a reflexive notion, and she implies activism, which also signifies the Marxist traditions of cultural studies. She has urged that the regarding of technoscientific objects as articulated is also an invitation to articulate new figures, like the cyborg, that cross-cut established boundaries as man/female or human/machine (Haraway 1997, 269).

So what kind of approach to the study of novelty and technological objects do articulations suggest?

First, articulations are not limited to expressions or labels used by actors in a specific context; rather, they are conceptual delineations, like class, identity, or habitus. This urges one to be clear about novelty, too. Understood as an articulation, novelty is not limited to what scientists, artists, or an audience regards and labels as better, progressive, aesthetic, of generally new. It is rather a *conceptual account* of situations with which one can characterize phenomena and processes. Hence, a study of novelty as articulation is one that searches for temporal qualities in the diverse compositions that signify the becoming of technological objects.

Secondly, the main issue of delineating those qualities is *coherence*. Articulation as heuristic focuses on the temporal coherence of combinations, which is not determined but is still structured and directed. From Hall's perspective, this would be the sense of unity that occurs at a certain point in discourses of power, and, from Latour's perspective, this would be when propositions are rendered as entities. In this regard, technological objects articulate their novelty in moments when elements click in and create meaningful unity. In contrast to articulations of class or habitus, novelty is a sense of unity that goes beyond previously existing things – that embodies difference.

Thirdly, articulations emphasize the *heterogeneity of elements* that connect in technological objects. It is a heuristic device for drawing attention to the ways elements from diverse realms such as signs, materialities, and bodies connect and appear unified. Furthermore, it stresses how structural elements and immediate situations are related in technological objects. This urges one to pre-set certain heuristic elements under articulations in order to broaden the heuristic scope and allowing the comparison of different articulations. In the following, I sketch three elements that articulate technological objects and their novelty: figures, technicity, and enactments. Each of them is concerned with different modes in which technological objects become part of a shared reality.

1.4.2 Figures

The first element that I regard as articulated are figures. Figures are related to what I have so far called semiotics[10], but are somewhat more specific, as they stress the continuing individuated forms that are built in and from stories of technological and scientific progress. Historian of science and feminist scholar Donna J. Haraway prominently discussed the structuring efficacy of figures and

---

10 When I speak of *semiotics*, I refer to its very basal meaning as the study of signs and symbols and their uses and interpretations. I do not implicitly refer to a particular semiotic theory unless I mention it as such. This basal understanding of terms that are conceptually declined elsewhere also includes related terms like *signifier*, which I understand as a symbol, sound, or image that represents or points toward an underlying concept or meaning.

1.4 Toward Novelty as Articulation

their embodiment within technoscientific discourses. By accounting for figures, Haraway has stressed, for instance, how figures of Christian realism are repeated in the technoscientific sense of history and progress in the United States (Haraway 1997). She understands figures as topics and tropes, articulated and drawn as categories of existence. A figure can be equally understood as a constitutive concept for, as well as the semiotic and material effect of, specific practices (Haraway 1997, 11):

"Figures do not have to be representational or mimetic, but they have to be tropic; that is, they cannot be literal and self-identical."

Haraway's perspective focuses on the power of language to relate technological objects and meanings to one another. It is less concerned with pragmatic use and ways technologies work as differential machines that produce material causalities. Rather, it stresses how difference is made by relating technologies to the specific *stories* of their existence. These stories are not arbitrary or limited to a specific object, but are part of shared imaginaries and discourses. They refer to activity and the active reproduction and signification of objects via speech, text, and pictures. The term *imaginaries*, just like figures, has a more long-term connotation and, in contrast to stories, refers to continuing structures of signs and meaning. Imaginaries share with the more colloquial term "imagination" an evocation of vision and fantasy, but instead of referring to individual experience, imaginaries consider specific cultural and historic resources that the world makes available to us, according to our particular location within it (Suchman 2007, 1; cf. Knoblauch 2011). Haraway's argument focuses on the *tropes* used in stories that relate imaginaries and technological objects. Tropes are figurative and metaphorical elements, not only in the sense that particular terms and expressions make a story more florid, but also that an object's mode of existence is a matter of invoking associations across diverse realms of meaning and practice. *Discourses*, in this regard, are not what is said, but are those which constrain and enable what can be said; their practices define what counts as meaningful statements (Barad 2007, 146). A figure embodies shared imaginaries and is repeated in stories that inhabit discourses, scientists, authors, and audiences; it "collects up the people" (Haraway 1997, 23). Technologies, in that sense, are materialized figurations (Haraway 1997, 23), as they bring together material and meaning into a more or less stable arrangement (Suchman 2007, 227).

Robotics is a scientific realm whose discursive structure is vividly inhabited by figurative accounts (Suchman 2007; Suchman 2011a; Alač 2009; Šabanović 2014; Castañeda and Suchman 2014). In particular, the "human" figure structures how humanoid machines gain significance. Stressing the human figure is asking what it means to be humanlike in robotics and how the boundary between human and machine is drawn. For instance, Lucy Suchman identified three elements of humanness in contemporary artificial intelligence and robotics: embodiment, emotion, and sociability. Sociability is, for example, expressed through affective interactional encounters with a robot, like in the case of Kizmet, which was designed to closely resemble a child's expressivity. Through the enacting of a culturally shared meaning of the category "child," the robot is signified as a machine in its becoming – one that can be encountered playfully and whose technological adolescence might not yet have arrived, but is about to (Suchman 2007, 226 ff.). Robotics and artificial intelligence reiterate such culturally specific imaginaries and enact machines as embodiments of their figures. This is not to regard the humanoid as a scientific model of the human, but to make evident how roboticists imagine humanness and how they enact and make intelligible the similarities and differences of their robots.

The example shows how technological artifacts are signified through their relations to culturally shared categories of existence. In Haraway's writings, another aspect of figures becomes evident, namely, their *individuation* (Haraway 1991; Haraway 1997). For example, she has delineated the OncoMouse™ as a figure within salvation stories of the biotechnical war on cancer as the embodied promise for the cure. She/it is a living commodity, a genetically modified laboratory mouse patented by the chemical company DuPont. The stories that surround its existence places OncoMouse™ as a figure of a culturally privileged kind of salvation that rebuilds the world from a laboratory. Haraway has delineated OncoMouse™'s existence as an embodied mystery in a secularized salvation history of scientific knowledge, progress, democracy, and economic power (Haraway 1997, 85). These relations, which are distinct from those of a mouse as a creature of nature, individuate the figure OncoMouse™ as an entity related to culturally shared desires, values, and promises. It individuates through its specific history of origin, which is peculiar but seductively auguring.

However, the story of OncoMouse™ is not merely harmonic as if its meaning is definite and stable. It is also a story of *contention*, which negotiates ethics, patenting, and the role of science. This contestability is crucial to Haraway's understanding of figures. She has stressed that figures are not simply inscribed into processes, but also trouble them. A figure must "involve some kind of displacement that can trouble identifications and certainties" (Haraway 1997, 11). In artificial intelligence, for instance, the discourse not only considers human intelligence as an ideal for the machine, but also contests what intelligence means, as computers have destabilized boundaries between human intelligence and forms that are more primitive. Katherine Hayles has spoken of "computing the human," not in the sense of inscribing human properties into machines, but rather of shaping our notions of the human. Human and machine are no longer measures of each other wherein one falls behind the other according to its capacities, rather computing the human stresses that the human cannot be adequately understood without ranging it alongside the "intelligent" machine(Hayles 2005, 148). Hayles' example shows how the "human" figure in artificial intelligence individuates because its intelligence is a benchmark, but, equally, the figure enables the contesting of what human intelligence actually is.

The duality of individuating and contesting is related to another aspect of Haraway's figures. The examples mentioned above stress the agency of language, but also implicitly stress a figure's *embodiment*. Emphasizing the role of bodies is of central concern for feminist research. This regards questions such as how actual bodies repeat cultural perceptions of the female ideal and how embodiments enforce boundaries and isomorphism. The embodiment of figures is also a relevant aspect of the ways technologies become part of a shared reality. The entailment of technical embodiment and mystified figures is shown in Suchman's study on "subject objects." On three demonstration sites, she observed how humanoid robots become models of the future. Unlike commercial demonstrations, which prove the functionalities and purposes of new technological artifacts, demos of robots are a matter of their existence and encounters of a cultural form of the uncanny, with something that is similar yet still different. Technical and popular imaginaries mobilize the figure of the robot not in its actuality, but in its potentials. In local encounters, these travelling stories and material assemblages interact and render the humanoid as "never quite realizing its promise but always also exceeding the narratives that animate it" (Suchman 2011, 133). In this sense, embodiment is a matter of a figure's existence, which entails specific dynamics triggered by physical encounters. A figure is materially embodied, but not determinedly bound to a specific material.

In the preceding paragraphs, I have outlined four aspects of Haraway's understanding of figures: the power of tropic language to signify technological modes of existence; the discursive tendency to individuate figures; how this individuation enables contentment, which is fundamental to a figure's existence; how figures are differently encountered through their embodiment. These four aspects are an exegesis, and other readers might have highlighted other aspects according to their own questions. I regard these aspects as crucial to how figures enable or prevent articulations of novelty. They stress semiotics of novelty and progress, how language individuates to mark objects as distinguishably different, and how language and embodiment might destabilize relations – all these aspects can structure how novelty is articulated. What is considerably different for this figure-perspective, as opposed to other theories that account for the relation between discursive practices and novelty, is its not starting from institutional settings and their conventions (cf. Polanyi 1962; Becker 1982). Of course, figures are significantly related to institutionalized fields like science and art, but the articulation of an object as new might also signify its existence in relation to cultural imaginaries that lie across different institutional realms. This could be, for instance, the human figure that is articulated in science and in art with distinct similarities and distinct differences. Thinking through figures allows accounting for the convergence of referencing, without negating the differences of specific settings.

1.4.3 Technicity

Technicity is the second heuristic element that I place under articulation. In contrast to the semiotic connotation of figures, technicity refers to the material dimensions of technological objects, their technical character, and the structuring dynamics that stem from assembling technical parts. In order to elaborate on the notion of technicity, I mainly refer to the philosopher of technology Gilbert Simondon. His ideas have recently been developed in philosophy, stressing what his thinking has meant for contemporary technologies (Hörl 2008; De Broever et al. 2012; Hoel and Tuin 2012; Iliadis 2013), as well as to what extent his concepts can be adapted for building social theory (cf. Venn 2010). I would like to refer to Simondon in a different, more pragmatic way and develop technicity and ontogenesis as heuristics to empirically account for processes through which technological objects come into being (cf. Mackenzie 2001; Mackenzie 2005).

Simondon's philosophy about the becoming of technological objects was developed by him within a larger framework to explain individuation without presupposing "the existence of a principle of individuation that is anterior to the individuation itself" (Simondon 2009, 4). He developed his philosophy as distinct from substanstialism, which regards a being as consistent in its unity, and Aristotelian hylomorphism, which considers the individual as the coming together of active forms and passive matter. Ontogenesis and individuation are the starting points from where he unrolled his arguments and concepts. *Ontogenesis* is the more general of the two terms, referring to a "becoming of being," comparable to its biological understanding, where the term refers to the origination and development of an organism (Barthélémy 2012, 219). *Individuation* is part of that process as the genesis of the individual, in the sense of a thing or an entity that can be considered different from others. Whereas the clear distinction of both terms is not of importance here, it is crucial to stress that it is *genesis* that concerns Simondon's thinking the most. He wanted to understand "*the individual through the individuation, rather than the individuation through the individual*" (Simondon 2009, 5, author's emphasis). Thus, he reversed individuation by prioritizing the operations from which the individual comes to exist and which of its characteristics reflect the development, the regime, and finally the modalities of

1.4 Toward Novelty as Articulation

its existence. He thought of individuation as a continuous becoming, and of the individual, or what is perceived as an entity, not as an endpoint, but as a situated and temporal form that always deviates from a hypothetical finality (Simondon 2009, 5):

> "The individual would then be grasped as a relative reality, a certain phase of being that supposes a pre-individual reality, and that, even after individuation, does not exist on its own, because individuation does not exhaust with one stroke the potentials of pre-individual reality."

This pre-individual reality is characterized by what Simondon called *metastability*, which causes the continuous flux of existence. Whereas stable equilibriums are in a state of low energy as all possible transformations are realized, metastability is a state that is charged with potentials. In metastable states, slight modifications can provoke chains of activities that break order and alter a system's state of being. These activities or structuring flows are what Simondon called *transduction*. It is an operation that propagates from one element to the next (Simondon 2009, 11). As such, transduction is a way of thinking about how individuation depends on not-yet-structured potentials or pre-individual states. From a metastable state, it is an ontogenetic operation that provisionally resolves incompatibilities between different orders or different zones of a domain, and leads to individuated beings, such as things, gadgets, organisms, machines, or the self (Mackenzie 2005, 393). Simondon's favorite example of transducing individuation in a metastable system is crystallization. A saturated solution is a metastable system, as it only takes a seed to break the temporal order of the liquid. The interference causes the potential energy of the solution to propagate from one element to the next and transform into crystals. Each constituted molecular layer serves as an organizing basis for the layer currently being formed. This transduction entails the transfer of energy within the solution and its simultaneous structuration and modification. Hence, transduction indicates the meeting of two disparate realms and the beginning of the process of individuation. It might be described as the process through which different properties interact among each other to produce something that is ontologically new (Iliadis 2013, 12-3).

Whereas the preceding terms – individuation, metastability, and transduction – are part of Simondon's general philosophy, I would like to continue now with notions directed toward the existence of technological objects.[11] These notions are embedded in individuation and transduction, but Simondon developed them in accordance with technological genesis. *Concretization* is the pivot-

---

11 I would like to emphasise that Simondon's thinking is characterised by the symmetry of terms used for the psychic, collective/social, natural, and technological realm.

al concept of these notions (Simondon [1958] 2012a, 19). Like individuation, concretization is a process and not a final state. In that sense, technical objects are never absolutely concrete, but are striving for and equally deviating from their pre-individual potentials. Their genesis is a convergent process in which the object "pulls in" the assembled elements; it is a movement towards the essence of the technical object. Concretization is not to be confused with the materialization of a concept and, respectively, also not to be confused with knowledge processes that communicatively institutionalize an object. It is the genesis of a technical object, wherein the object is not a cognitive thing advancing its becoming, but what is present in every stage of becoming: "the technical object is unity as a unity of becoming" (Simondon [1958] 2012a, 20).

There are two aspects that characterize what Simondon regarded as the technical becoming, which he called concretization (cf. Barthélémy 2012, 208-9). The first is the *internal resonance* of the elements that compose an object. It is a fitting-together of physical components that fosters a growing organicity. Each piece fits into the functional unit, like transducing crystals, whose structuring follows internal causalities. The second aspect is the *pluri-functionality* of an element. Instead of having a single function, concretization entails the integration of the redundant properties of each element. As an example, Simondon referred to different cooling systems in combustion engines. Water-cooling is not concrete in his sense, as the cooling system requires a sub-ensemble of tubes, pumps, liquids, etc., which are external to the functional unit of the engine. In contrast, air-cooling does not rely on an external radiator, but is based on cooling rips that are part of the cylinder pistons. The pistons are pluri-functional, as they not only provide a physical structure through which the cylinder determinately moves, but also channel the airflow to prevent the piston from overheating. In that sense, air-cooling is more concrete, because cooling does not require an additional sub-ensemble for/in the functional unit of the engine, but is achieved by integrating the redundant characteristics of assembled elements. From that perspective, redundant properties are not necessarily problematic, but contain potentials that become effective through concretization.

Simondon conceptualized concretization as a convergent movement to explicate internal processes and causalities through which objects individuate as units. That is, concretization is a movement that sharpens the technical character, or the *technicity*, of an object. This is exactly the opposite of an object, whose genesis is an adaption to the environment, in the sense that it fits a given milieu. Original to Simondon's thinking is that he regarded concretization as a process that tends toward an *open machine*, as opposed to adapting, which dead-ends technical development (Simondon [1958] 2012a, 47ff.). Adaptation regards the modification of an object as fitting the material and human conditions of its us-

1.4 Toward Novelty as Articulation

age, or an object splitting into modules (i.e. software/hardware). Both cases entail adapting an object to a given technical (i.e. electricity) or spatial (i.e. railroad construction) milieu. A concretized object defies such adaption. Concretization is not a process conditioned by a given environment, but a process that conditions a new "associated milieu." That is not to say that external processes do not matter. On the contrary, it means that technical objects have the potential to mediate between two disparate milieus and, from there, to develop a new associated milieu through concretization. Simondon's most famous example for that process is the Guimbal turbine. The Guimbal turbine is a tidal power plant that combines a turbine and generator into one unit. Its specific engineering is based on the plurifunctionality of oil and water. The water not only carries energy from tidal movements into the turbine, but also carries heat away from it. The oil not only lubricates the generator, but also prevents water from infiltrating. At this point, water and oil are carrying specific potentials that are mediated through the technical object. Nevertheless, the process goes further in the case of the Guimbal turbine, as oil and water create a new self-sustaining milieu. The potentials in the oil and water have interlinked in such a way as to regulate the transfer of energy into the turbine and of heat out of the turbine automatically. By interlinking water and oil so that their potentials are fully integrated, which re-arranges the border between external water and turbine, the turbine increases its autonomy, as it operates without external maintenance. This example shows how technical objects may interweave different realms and create new relations, not by adapting to external factors, but by integrating potentials into a functional unit. This is what Simondon regards as an "open machine." As opposed to automation, which mimics human behavior, an open machine conceals a margin of indeterminacy, which makes it sensitive to external information (Simondon [1958] 2012a, 11).

For Simondon, concretization and integrating an associated milieu are not only technical improvements, but also elementary conditions of progress. He wrote that the necessity not only for adapting to a given milieu, but also for relating different realms and integrating these into the development of a functional unit increase autonomy and concretization. "Here lies true technical progress" (Simondon 2012, 50). In contrast, adaptation always refers to something that has existed prior; as such, it always runs after its conditions instead of affecting them and causing new ones. This idea of progress is built on a specific kind of innovative acting. In order to create a self-sustaining technical system, one needs inventive anticipation that does not draw upon relations found in nature or in other technical objects, but that is directed toward realizing systematic convergence. Simondon emotionally wrote that "it is an act of life" to go beyond the given reality and its current system toward new forms that only exist as they constitute a new system; "if a new organ appears in the course of evolution, it may only be maintained if it realizes self-sustaining convergence" (Simondon [1958] 2012a, 52).

1.4.4 Enactment

The first two heuristic elements, which I regard as articulated through technological objects, emphasize structural dynamics that go beyond individual actions: figures emphasize the continuity of symbols and signs within technoscientific imaginaries, and technicity understands technological development as convergence guided by material characteristics. In contrast, the third element emphasizes the temporal and situated performance of objects – the *acting out* of objects. I address *enactment* as a discrete heuristic element of articulations so as to give space to the contingencies and situated practices of how figures and technicity come to matter. In enactments, figures and technicity meet and interweave with materialities, stories, and bodies. Whereas articulation is the temporal sense of unity in the genesis of technological objects, enactments are the activity of articulating, as they stress the bringing together of elements. Addressing them as a heuristic element that is part of articulation puts emphasis on the tangible and temporal situatedness of action, which beholds dynamics that figures and technicity do not address.

Much of this heuristic element is inspired by the literature discussed as the perspective of novelty as differential pattern. From that perspective, objects are produced as distinguishable units through the temporality of interaction (Dewey 1938; Pickering 1995; Rammert 1999; Barad 2007). They are performed through the circularity of interactions that mutually constitute an object and subject's temporal modes of existence. Karen Barad has emphasized in her agential realist approach the enactment of boundaries that render entities visible. Boundaries are not a matter of distinctive categories that are given *a priori*, but are actively produced in an alliance between humans, instruments, discourses, and contingent agencies in the world under investigation (Barad 2007). This perspective not only is materially relevant, but also entails morality, power, and valorizing distinctions. Distinctions such as, for instance, "art object" versus "scientific object" or "old" versus "new" are not fixed properties, but relations that need to be enacted and that take their situational circumstances into account. This entails not regarding the RBO Hand and *Mirage* as fixed samples of two institutionalized units in a field. This is not to say that their contextualization does not matter. On the contrary, it means acknowledging the variety and situatedness of relating and referencing. Enactments stress the consideration of novelty and difference as repetitively reproduced in resonance with an object's temporal and spatial location.

In this regard, situatedness has a twofold meaning. First, it is related to the term "situated actions" or "practice" as understood by Suchman (Suchman 1987). She introduces the term so as to underscore that every turn of action depends on its social and material circumstances. Rather than considering actions as based on a rational plan that leads linearly to certain conduct, situated actions consider how people use their circumstances to achieve intelligent action. This understanding of action exists very much in the pragmatist sense, which also considers the circularity of problem and action that translates contradicting situations into a temporal state of unity (Dewey 1938). Secondly, the situatedness of enactments also entails a notion that might also be regarded as interactional framing. Framings consider the sociality of situations produced through interactions. This is, for instance, how specific gestures, ways of talking, or bodily positions signify the shared meaning of situations (Goodwin 2000). Wally Smith investigated, for instance, how presentations of technologies are interactionally framed and re-framed over the course of demonstrating new functionalities (Smith 2009). Framings consider primary frames, which are rather stable and set in advance, like "presentation," as well as more contingent framings and re-framings within interactional sequences that occur, for instance, when something unexpected, such as failing technology, happens.

In her study on protein modeling, Myers considered both meanings of situatedness: as situated action and as framing (Myers 2008). She placed emphasis on the role of the body in enacting protein models. In the first meaning of situatedness, the body is an epistemic tool that researchers use to make the abstract models present within their epistemic work. By stressing the role of the body, Myers wanted to capture the interrelations of mental and physical models that are not to be regarded as dichotomies. In a similar vein, Alač has stressed how roboticists use their own bodies to make present the inner workings of a humanoid robot (Alač 2009). In terms of the second meaning of situatedness, Myers investigates the role of embodiment in public presentations. As such, researchers use their bodies to make graphical objects tangible, and they employ gestures and movements in communication with novices in order to flesh out and relay their knowledge about otherwise only virtual objects (Myers 2008, 180).

In this respect, I consider situatedness as the immediate context of enactments and not only as the historical and institutional setting. Enactments consider how differently objects are put into action in diverse settings – for instance, how engineers perform the workings of a prototype that is far from realization, in contrast to how research papers enact that same object. Accordingly, enactments stress how diverse elements are selected, translated, and articulated as novelty. Furthermore, they also stress how connections to other objects are made in order to distinguish the novelty of an object. This includes, for instance, references and their dependence on a specific framing, setting, and discursive practice.

1.5 Study Design II: Abstraction, Critique, and Comparison as Method

The theories that I collect here to inform what I consider articulations of novelty formulate heuristics on a high level of *abstraction*. Authors like Haraway and Simondon have written philosophies of technology and scientific process that seek to delineate the foundations of structure and movement, or matter and meaning. They are close to my empirical investigation of the RBO Hand and *Mirage*, as they are concerned with the ways technologies exist, but, in terms of abstraction, their philosophies operate on a much higher level than my ethnographies. Nevertheless, I have chosen such abstract heuristics on purpose. Abstraction allows one to make connections between objects that appear to be very distant from an empirical perspective. It allows one to see similarities between processes that happen in diverse settings, but that also share certain dynamics. In this regard, articulations entail the *ideal types* of figures and technicity, but my delineation of them does not attempt to be as abstract. Rather, articulations are at a level of abstraction between empirical observation and universal theory. They are the relationship between what is apprehended and what seems to demand apprehension; between observing and understanding (cf. Strathern 1999, 9). Articulations connect what I want to understand about novelty and what I observe at a particular place in a specific time.

Besides abstraction, another reason for choosing these theories is that each of them approaches my central tension between relating and individuating – but each does so differently. Figures ask how stories and narratives reiterate categories of existence, technicity stresses how material capacities converge in technological objects, and enactments focus on how interactions create a sense of unity. Hence, they are all concerned with what I regard as articulation, but each would answer the question of novelty differently. Potentially differing answers are also potentials of a heuristic as such, as they give room for answers from a wide spectrum and prevent redundancy. With that said, I can express my central research question in a condensed way and ask: *how do technological objects articulate novelty?* This is the central question of this study.

However, discussing abstraction also raises the issue of *critique*. I have already outlined that considering novelty as articulation is an attribution that seeks to delineate certain qualitative shifts in an object's becoming. This is very different from regarding novelty as a label used in the empirical field. Considering novelty as a label assigned by actors to promote their technologies as better or superior lines up "novelty" beside field semantics as "innovation" or "progress," which carry the promises of salvation of technological progress. From Haraway's perspective, these are technoscientific tropes charged with values and promises. This is certainly an attractive argument, and I want to use her critical view, but this perspective alone leaves little room for considering novelty as a specific quality. This does not mean regarding novelty as something "better" like actors do in the field, but rather allowing a conception of novelty that is more than semantics enacting a difference from whatever came before. This is certainly the point for Simondon to enter the argument. His philosophy is concerned with how objects come into being and individuate. Besides this abstract starting point, his theory of concretization contains aspects through which an object becomes qualitatively new. These qualities are, for instance, the tipping point when different material realms interlink and form a new associated milieu, or the quality of a concrete object integrating redundant characteristics from each technical element. These specific technological qualities not only mark an object as different from others, but also characterize moments when potentials unfold and create a distinct kind of efficacy. In my view, this is a unique argument, and Simondon was a thinker who provided a promising terminology. Nevertheless, he lacked what Haraway has – an understanding of discourses and the modes of existence within them. His perspective does not regard the histories that are attached to specific kinds of objects or that structure the ways objects are enacted as novel. In this regard, articulations are a matter of combining critical epistemologies that do not take actors' constructions of novelty for granted, with heuristics that allow the delineation of qualitative shifts in an object's becoming and the material and semiotic compositions it articulates.

Discussing the theoretical heuristic of this study, including what abstraction implies, leads to a problematizing *comparison*. Proposing a theoretical heuristic implies that the study's approach cannot be a traditional ethnography – this would require a more open analysis. Hence, the comparative study design here is somewhat in between the thick empirical descriptions of ethnographies and the more theory-driven objective of learning something beyond the empirical case. This methodological paradox is not new as such and has lately been matter of discussion, to which I have also contributed a methodological paper on the two cases analyzed here (Stubbe 2015). In that paper, I developed the comparative approach that I also follow throughout this study. The methodological paradox is that, on the one hand, science and technology studies usually argue from the ethnographic perspective that contexts exhibit different and incommensurable traits, whereas, on the other hand, comparison involves the investigation of discrete contexts to elucidate their similarities and differences (cf. Strathern 1991; Strathern 1999; Niewöhner and Scheffer 2010; Jensen et al. 2011; Morita 2014). . In the paper, I broke down the discussion to three entangled issues, which I reconfigured based on engagements with my own data and Marilyn Strathern's approach to analogical comparison. The three issues are: the construction of comparability, the perspective from which something is compared, and one's own bodily involvement. In the following, I briefly reiterate some conclusions that sketch what kind of comparison this discussion proposes.

The first issue concerns the *construction of comparability*. Objects of comparison are not "out there" as predetermined units, but are produced through the research process. Regarding the construction of comparability, Strathern has written that it is the act of comparison that constitutes relationships, not intrinsic qualities of phenomena. As mentioned concerning the sampling of the RBO Hand and *Mirage*, they are not compared as already institutionalized objects. It is left to the research process to determine whether particular articulations can be delineated that make an object identifiable as one of art or science – and, more crucially here, as a novel object. Such an approach to comparison does not assimilate two phenomena into a deductive category, but draws lines between discrete phenomena in order to delineate their sameness and difference, as well as to give image to the continuities that exist across the complexity of situations. In Strathern's work, continuities are not articulated as questions of homogeneity, but as "proximities in space and time" (Strathern 1991, 55). Heuristics as articulation form a kind of integrated circuit between parts that work as significant continuities that can be delineated as novelty.

The second issue stresses the *perspective from which the comparison is made*. Sociologist Joachim Matthes urged the consideration of comparison as a cultural operation that calls upon the researcher to step back and ask what constitutes the experience of sameness and difference (Matthes 1992). The experience of alteration is not substantive, but relational – including how the researcher relates to what is considered different. To articulate her relation to the subject matter, Strathern made use of Haraway's cyborg figure (Strathern 1991, 54). Haraway later articulated her own research perspective by figuring a "modest witness" (Haraway 1997). Her modest witness is not oppositional to its subject matter, but implicated and literate, just as it is suspicious and worried. It is inside the "Net" of stories and agencies and simultaneously learns to avoid its narratives and realities (Haraway 1997, 3). In this sense, I go beyond the position of a silent ethnographer and instead engage with the actors. I ask about how things work and take part in conversations; I take the liberty of re-arranging observations and assembling these into narratives according to my interest in novelty; I re-arrange what I see and enter into dialogue with the literature that I am familiar with. This practice allows for going beyond description in order to identify continuities in two empirically distant cases. A modest perspective enacts connections that respond to one's own agency within situations just as it may render continuities between distant locations.

The third issue addresses one's own *bodily involvement* in situations. A researcher's agency within situations not only matters in terms of intellectual perspective, but also in terms of bodily engagement in situated enactments. In methodological terms, this urges the implication of oneself in the situatedness of practices that are simultaneously witnessed. Empirical instances of this study entail how Deimel and Baecker react to my ethnographic engagement in their practices. They do so with gestures and rhetorical use of imperatives in their stories. If an object is enacted as new in such situations, it is referenced as different in anticipation of my response. In this sense, I am figured through the situation, just as my perception figures the subsequent account of what happens. This might be, for instance, when bodily movements enact objects according to my presence in the studio and lab, or when my bodily presence in a crowd of spectators influences the following account of public enactments. A modest witness who is making a comparison needs to consider and use one's own bodily and intellectual position in the circuits of materialities, stories, and bodies that articulate an object's novelty.

These issues address the methodological paradox of comparison from a perspective of articulations. This paradox is how to account for the situatedness of events while identifying continuities to define abstract processes. My strategy for solving this problem entails the pre-setting of heuristic elements, such as figures, technicity, and enactments, as well as the openness to drawing inductive connections between cases. By contextualizing what is found in one location with what is observed in another and relating these to what is already understood concerning a particular question, I identify and name processes that are significant and accountable across sites. This comparison connects the different temporalities of encounters, changing literacies, and the flux of questions and contingencies. Its witness is local, as it is immersed in the net of materialities, stories, and bodies, just as it moves on and re-arranges what is understood and what needs to be understood.

1.6 Overview of the Study

The study is structured in three parts, of which the first is the preceding introduction. The second part is the comparison of the RBO Hand's and *Mirage*'s becoming. That part includes three chapters, which each capture one articulation of novelty. These articulations are: identity (Chapter 2), form (Chapter 3), and difference (Chapter 4). The study's third part includes discussions and conclusions that contextualize the findings as well as summarize their main issues.

The comparison of the RBO Hand and *Mirage* begins in Chapter 2 in a state wherein both objects have not yet been fully realized. There are stories about the new project's origins, traces of tinkering processes, loose imaginaries, and figures, as well as embodied ways to flesh out what the future object might be like. However, there is no natural belongingness or force that dictates how these elements are connected. Rather, coherence is an accomplishment in this state that requires activity and effort from the actors. The chapter has its entry point to this accomplishment in following "ideas." I understand ideas not as cognitive plans or intents here, but as actors' categories within stories that signify future objects. They are words and labels that *do* something to the object-character of the otherwise loose elements: ideas connect selected past events into a coherent storyline of an object's origins, they refer to imaginaries of larger significance, and they enact a prototype's potentials. Analogously to Mead's theory of the self, I delineate the articulation of diverse elements in this state as the building of an object identity. This identity signifies an object's biographical trajectory; it contains generalized accounts of other objects and the collective imaginaries that give reason for an object's oppositional character. In this sense, ideas are necessary for articulating coherence in the diversity of materialities, stories, and bodies that make an object. Novelty as an object identity connects diverse and previously loose elements to give body to the sense of unity that transduces their situated enactments.

Chapter 3 takes a presumably oppositional stance: its main issue is how technical forms evolve that articulate a distinct kind of physical efficacy. Instead of creating coherence by acting out an object's potentials, as done in the previous chapter, this chapter focuses on how potentials concretize into functioning technical units. The novelty of a technical form, which embodies a distinct kind of working, needs to articulate several structural demands. Firstly, actors need to consider certain functionalities that constitute a type of object. For the RBO Hand, this is grasping, which demands finger-like elements to bend and capture items; without that functionality, the Hand will not be a robotic hand. Similarly, *Mirage* is a cybernetic machine in terms of its capacity to convert environmental signals into a contingent and dynamically adaptive output. Secondly, a form needs to articulate the hybrid constellation in which it is produced and exists as a working device. This issue stresses that technical functionality is distributed among different agents: infrastructures, materialities with varying capacities, mechanics, algorithms, and humans who conduct the constellation. The distributed character of technical efficacy brings up the critical subject of how one can speak of novel forms when their efficacy depends on several other agents. Hence, delineating the novelty of form requires the characterization of the relation between the new form and the constellation of its existence. Comparing the RBO Hand and *Mirage* brings up two different articulations: the RBO Hand builds upon the constellation of its existence, which allows the simplicity of its form, whereas *Mirage* incorporates the agents of its constellation, which increases the form's complexity. Articulation of a technical form stresses how heterogeneous elements come together and connect and how novelty becomes a moment when parts click in and work together as a technical unit.

Chapter 4 moves into the discursive arenas of robotics and media art so as to capture how both objects' differences are valorized as novelty. Nevertheless, the comparison has its beginning, once more, in the robotics laboratory and studio. The discursive enactment of both objects is structured by shared discursive practices, which influence how the RBO Hand and *Mirage* are translated into other entities that make their distinct characteristics travel beyond the immediate context of their production. For instance, the RBO Hand's distinct grasping style translates into a graphical representation that fixes the results from a scripted experiment that takes place in the laboratory. The central activity for Deimel and Baecker to make their objects' meaning intelligible is referencing. Whereas the RBO Hand's references are conceptual as well as figurative, *Mirage* is accompanied by associative references that point out peculiar aesthetic similarities between scientific technologies. The categories, which the references signify, are reiterated in the discursive recognition of both objects. These categories are "compliance," which marks the RBO Hand's distinct grasping style, and "the hallucinating machine," which is the pivotal trope of *Mirage*'s significance. The recognition and reiteration of these categories entails citations by other scientists, exhibition catalogues, public articles, and awards and prizes, all of which mark both objects as valuable contributions to robotics and media art. The central issue of novelty as difference is that difference is not just the relation between one object and a group of other objects; it also requires the stabilization of categories that embody that difference. These categories are the focal points of discursive valorizations and also become those elements of an object that fuel further inquiries into novelty from other actors.

Chapter 5 begins Part III by discussing and contextualizing the findings in light of the notion "aesthetic reflexivity." The chapter builds upon the study's findings and attempts to broaden its scope by connecting certain aspects to the more general discussion concerning reflexive modes of human-technology engagement in postmodern societies. In so doing, the discussion jumps back to the initial "simple, but intriguing similarity" and delineates how far the deep material engagement of technology in science and art signifies a counter-action to the artificial purification of scientific models and abstract theory. These approaches can also be found in other science and art projects that draw on implementing materials instead of theoretical statements and, hence, reconfigure human-technology-constellations. This kind of reflexivity builds upon aesthetics, effect, experience, and *self-interpretation* instead of technocratic self-monitoring. From this critical stance, novelty opposes technoscientific images of progress and innovation, as it becomes a matter of experiencing and relating to the heterogeneity of semiotics, literacies, and materialities that are (re-)articulated through technological objects.

Chapter 6 is the concluding chapter of the study. It begins by answering the study's main question and sharpens up novelty's central tension between individuating and relating objects. The conclusions include a summary of the comparison's findings as well as a discussion of articulations as a fruitful heuristic to study the relation between novelty and technological objects. The study ends with an outlook on future issues concerning the study of technology in contemporary societies.

PART II
Analysis – Three Articulations of Novelty

2 Identity: How Loose Elements are Connected

This chapter delineates novelty as the articulation of an object identity. It attempts to delineate how a sense of unity evolves in a state prior to an object's full technical realization, when elements are loose and technically unconnected. In a certain sense, the chapter addresses the ideas of an object. However, it is not about thinking, and my argumentation takes a different perspective than that of regarding novelty as an invention based on ideas. I do not attempt to look inside Deimel's and Baecker's heads and give an account of how they thought of ideas, plans, or genius inspirations for their objects. In this respect, I agree with the natural suspicion of sociologists towards processes that are supposed to happen in minds alone. Nevertheless, it is not only disciplinary provenience that keeps me from thinking of an idea as something cognitive; what I have observed pushes the workings of the mind into a peripheral position. Surely, both creators are clever, creative, and highly technologically knowledgeable, but that was already obvious before I started my investigation and before each of them attempted the project of a new object. Still, referring to an "idea" was so repetitive throughout all the conversations that it seems odd to neglect *what this term articulates*. In this respect, I argue that "ideas" articulate an object identity, as they bring together the histories and potentials of objects.

Referring to ideas is as repetitive throughout conversations as the term's meaning is diverse. For a social scientist, this poses a twofold problem. On the one hand, diverse connotations make it hard to pin down an idea as a particular meaning that is inscribed into an object (i.e. as a script), and, on the other hand, their ubiquitous reference makes it hard to ignore that ideas are somehow related to an object's becoming. Latour and Woolgar similarly described this ambivalence when they stressed how difficult it is to ship around thought processes while investigating "laboratory life" (Latour and Woolgar 1986). Thinking seems to be integral to the peculiar and mythical existence of scientists and their creational work, they write. Thus, sociological accounts can hardly keep ideas out of the analysis of an object's becoming. Instead of taking tales of inspiration and ideas for granted, Latour and Woolgar have taken the notion of someone having an "idea" as a condensed summary of a complex series of processes that have faded from the immediate situation. Aside from reconstructing the biographical path an idea has taken, which makes it unlikely for an idea to still be regarded as an individual act, Woolgar and Latour also have stressed considering the accounting practices that create and sustain thought processes (Latour and Woolgar 1986, 168).

However, their remarks remain rather short. Moreover, most science and technology studies seem to avoid speaking of ideas. Ideas seem to belong to the perspective of novelty as invention, which considers the creation of technology to be based on plans and intention – a paradigm that is thought of as overcome. Knorr-Cetina has reported that early laboratory studies from the 1970s and 1980s introduced the notion of practice precisely as a contrasting term to highlight that "one investigated scientists at work as opposed to the history of ideas, the structure of scientific theories, or the institutional settings of science" (Knorr-Cetina 2000, 9). The observation of practices and their collective and material ways of doing became a methodological trademark of laboratory studies, which differentiated their ethnomethodological approach from the history and philosophy of science. The notion of practice shifts the focus away from mental objects, such as the interest or intention that inform concepts of action, and toward the contingencies of collective knowledge production among heterogeneous agents. This shift seems to have swept ideas away from the vocabulary of science and technology scholars, too.

In this chapter, I do not attempt to work against the paradigmatic shift from histories of ideas toward practices. I retain a perspective that is concerned with the material contingencies of concepts (Pickering 1993), the "dirty work" of aligning ideas and problems (Fujimura 1987), the striving for the immutability of events (Latour 1987), or the locality of knowledge production (Knorr-Cetina 1988) – aspects stressed through the notion of practice. Thus, I do not open heads, but try to delineate what counts for an idea, how it is enacted, and what it articulates – or, to put it another way: what does an idea *do*? These questions focus on what ideas bring together and how they enact stories, materialities, and bodies. Hence, I go beyond Latour and Woolgar's proposition of stressing the trajectories that lead to an idea and eventually dissolve from the stories of invention, and instead avow ideas as significant for an object's novelty.

My heuristic begins by treating an idea as an actor's category. This addresses the stories told when someone speaks of an idea. Nevertheless, I go beyond semantics and do not reduce my account to verbal or written mention of the term "idea." Rather, I attempt to take an idea seriously as a category that is there for a reason – as an agent that does something. Hence, I follow ideas through diverse realms and situations in order to inquire about how diverse elements become coherent and distinguishable without an object's full technical realization. This account begins with the stories that have led to advancing the RBO Hand and *Mirage*, including the traces that inquiries leave behind in the laboratory and studio. Whereas the first section deals primarily with the selection and re-enactment of past events, the second part of this chapter deals with how ideas enact the potentials that are to become effective in the future. Here, ideas become tropic narratives and prototypical embodiments, as well as bodily performances. Closing this chapter, I conceptualize how ideas articulate an object identity. To do so, I step into dialogue with the theory of the self by George H. Mead, who regarded individuation as an interactional process based on responding to the attitudes of others.

2.1 Enacting the Past through Stories and Their Tangible Traces

This section is concerned with ideas as a matter of enacting the past. It begins with stories of deviation. Deviation occurred through experimental practice and stories retrospectively account for them as relevant for the new project. From these stories, the section moves on and sketches that deviation within material tinkering leaves traces in the laboratory and studio – but only those traces that matter.

2.1.1 Stories of Deviation

The RBO Hand and *Mirage* are not historical monoliths. They have a history – a story that is narrated as their origin. Unlike what Latour and Woolgar reported, that referring to an idea erases the path that had led to a thought regarded as original, the RBO Hand and *Mirage* have been accompanied by histories that enacted their trajectories. Such narrative enactments started with referring to an idea.

> "Such ideas develop over a longer period of time." (Brock, RBO Hand)

Although I had met Deimel several times, and he told me about the idea behind the hand, it became clear that the trajectory that eventually led him to build the Hand began earlier. Hence, I approached the director of the institute, Oliver Brock, and asked for an interview to get an impression of the Hand's origins. The director begins his story by stressing that such "ideas develop over a longer period of time." It all started in a previous position of his, when he worked with a colleague who was engaged in computer vision. Together, they thought about what kind of perception a robot would need for grasping an item. He referred to how the problem was usually understood in robotics – namely, as a matter of understanding the geometry of the item. In contrast, they came up with the "idea" that geometry might not be very relevant to actual grasping, but that an item's surface already indicates how it is supposed to be grasped. That means, Brock told me, that the interaction between object and hand leads to successful grasping. When a human hand grasps an item, its specific geometry is not crucial; human hands automatically adapt to an item's shape when they are closed. However, he said, these ideas are mainly about perception.[12] Therefore, he and his colleague moved on and developed basic perception primitives[13], which exploited the ability of robotic hands to adapt to items' shapes. Based on that experience, they agreed that it would make sense to let robotic hands do more work in order to improve a robot's grasping abilities. Such a reallocation of capacities could make the perception and planning of grasping tasks easier. The hand would gain "responsibility" in such an approach, as it would become more "competent." Brock continued and stressed that, in fact, every artificial hand has somewhat of a kind of compliance. Hence, they started to experiment by using standard robotic hands. Eventually, he saw a video of a research group that was building inflatable and flexible things from silicone. In that video, the research group showed how one of those things was able to pick up an egg. Brock recounted that this was the moment he thought, "That is the answer." Brock's and his colleague's following plan was to build a hand as competent as possible, which meant as compliant as possible. Silicone seemed to demonstrate a way that could be achieved: by giving the hand plenty of degrees of freedom and enabling the hand to control them – that is, the ability to comply with an item's shape by equalizing air pressure. Brock continued with his story, saying that, by accident, there was a colleague in the faculty who had actually worked for Disney Research, where researchers had already worked on molding and inflating silicone. Following that encounter, Brock went to Zurich and had a look at how the fellow researcher's group worked and how they fabricated different objects; he had a look at how the whole process worked. When he finally received his current position at the RBO Lab in Berlin, Brock advertised a position for somebody to conduct the project. He assembled all the materials and tools needed to mold silicone, including the first mold, which had a starfish shape. When Deimel arrived, he was immediately able to start working on a robot hand made of silicone.

The story of the RBO Hand, as told by Brock, began with the intersection of his experience from experiments in robot perception and his awareness of research activities that explore the capacities of soft materials like silicone. Connecting both elements, his experience from experiments and the potentials of silicone was necessary in order to come up with the "idea" of a soft hand for robotic grasping.

---

12 Perception is a research domain in robotics, comparable to grasping.
13 Primitives are developed in robotics as sub-routines or basic functionalities that can be combined with different applications.

2.1 Enacting the Past through Stories and Their Tangible Traces 77

Although the RBO Hand has been developed in a setting very different from *Mirage*, and its scientific provenience is clearly indicated in Brock's story, the stories of ideas still share similarities, as the following narrative shows.

"The finished object is always different from your initial idea. And you simply have to do something new to investigate it further." (Baecker, *Mirage*)

Baecker's history of *Mirage* has no clear beginning. Although there are continuous references that he mentions as ideas, it is hard for Baecker to account for them as starting points or beginnings for that specific installation. Rather, ideas follow him for a certain time, maybe even throughout several works. He might discard them once, but then picks them up again later. He told me that there is always a difference between an initial idea and the finished object. That difference causes him to build something new to investigate it further, as he said. Thus, an idea is not a strict plan, but has materialized several times as prototypes, test structures, simulations, etc. before it is referred to as the idea behind a piece. A common procedure for Baecker is rebuilding ancient technologies that he has read about and that are no longer easily available. There are several material traces of such attempts in his studio. He uses these leftovers to quickly build something new and see what a mechanical movement might look like so as to get a picture of its specific aesthetic.

In this regard, his story about the idea behind *Mirage* also starts with a technical episode. When he and I were sitting in front of his laptop in his studio during one of our first encounters, he showed me a simulation that he found online. The simulation was about sensors that monitor movements in order to predict subsequent steps. These basic algorithms are used, for instance, by short-term weather forecasters. Baecker comments on the patterns on the screen:

"But when I feedback, these peculiar patterns evolve. And the idea, the analogy is this dream story, that you build a machine that hallucinates. A machine that hallucinates, or learns, or perceives an image and suddenly starts to process and to hallucinate and then re-joins images."

I asked him whether he actively searched for these things. "Yes," he answered, he had looked at different things for a long time. In particular, he looked at artificial neural networks whose development applied dream analogies, too. He referred to Helmholtz machines, which are algorithms used to identify hidden structures in complex data sets. These machines are based on a "sleep-wake rhythm." After researching this idea for a while, he began to program something similar in order to see how it worked and what kind of patterns could be generated from it. For Baecker, approaching the idea of a "dream story" is to render visible the hidden processes of algorithms. He pointed to the screen of his laptop that was still showing the simulation displayed as a moving grid, and he said:

> "It is interesting how these movements evolve and in what direction each point moves next. I use such vectors to predict and generate subsequent images."

The story of the idea behind *Mirage* is that of an inquiry. The story enacts the idea as something abstract (a "dream story") that he had learned about through investigating technological concepts, and that he approached to make it comprehendible. This leads to a series of attempts in which idea and materialization commonly deviate. Baecker does not use the term "idea" to relate his inquiries to a specific installation, but states that ideas follow him. They might be discarded once, but can be picked up again and re-worked.

In this sense, both the stories of the RBO Hand and *Mirage* relate ideas to deviations. They are both stories of exploring technological capacities and accommodating resistance. In the two stories, the term "idea" connects past events, such as when something did not work as planned and deviated from expectations, and attempts to build something new or to explore something in more detail. The storytelling of both actors does not refer to ideas so as to indicate the ingenuity of a single person and, furthermore, does not erase the history of events that led to a new object, like Latour and Woolgar have stressed. On the contrary, these two stories use ideas to select specific past events as relevant for building a new future object.

2.1.2 Material Traces

Both Brock and Baecker indicate that deviation was mainly a result of material practice and observing the difference between expectations and results. In both cases, such differences triggered explorations to find solutions or new approaches to their specific problems. However, the histories of such explorations are not just verbally reproduced stories. There are material traces of such explorations, which do not vanish from the surroundings and places of an object's becoming, but remain as materializations of past events.

> "Surely, there is an idea in the background that drives you to move on in specific directions." (Deimel, RBO Hand)

During one of my first visits to the robotics laboratory, I also visited the laboratory's workshop. The workshop is mainly used to maintain the technological infrastructure. Although the institute does not primarily engage in hardware

2.1 Enacting the Past through Stories and Their Tangible Traces 79

development, and many of their research projects deal with the development of robotic algorithms that run on standard robotics hardware, the institute's workshop is comprehensively equipped. It is located in a large room with several tools, such as drills and electrician equipment. The infrastructure and tools needed to work with silicone and to manufacture the RBO Hand are also located in the workshop. When I was there, most of these tools were placed on one table, which left the impression that it was a designated place within the workshop where the material tinkering practices concerning the RBO Hand took place. The tools on and beside the table included, for instance, apparatuses like a vacuum chamber and a precision gram scale. Furthermore, there were smaller tools for handling silicone, like vinyl gloves (for handling liquid siliconees), polyethylene cups (for mixing siliconees), chopsticks (for mixing small amounts of silicone), and more general tools, like a cutter, scissors, a metal ruler, and a cutting mat. There were consumables like silicone, mold sealant, and sewing thread, too.[14] The tools were not left in a mess, but not in painstaking order either. The whole arrangement seemed to be in constant use.

Close to the tools, at another table, lay a primitive version of the RBO Hand (*Figure 5*). Actually, it was difficult to call that silicone shape a hand, as it was hardly possible to grasp items with it. However, the shape was "hand-like." It consisted of three parallel fingers and a palm. Their size relation was different from a human hand, as the palm was considerably smaller. This proportional difference between fingers and palm hindered the shape from capturing items in order to grasp and hold them. Although this silicone shape did not yet have the crucial capacity of a robotic hand, it was still significant to the RBO Hand's material development. Its significance is the basal functioning of its assembled materials. The shape's main body (fingers and palm) was cast from one piece of silicone, which is approximately 1 cm thick. An inflexible rubber layer was glued on one side of the shape. This layer makes the silicone bend inwards when inflated with air. The whole shape was wound with a thin thread, which prevents the silicone from simply blowing up and directs the air pressure toward a bending movement. Hence, the shape could already perform a capacity, which was of relevance for the forthcoming RBO Hand: it deterministically bends when inflated.

As I picked up the silicone shape to have a closer look, Deimel, who was accompanying me, told me about how they approached the material to use for grasping. He emphasized that silicone is a material that is rather easy to appropriate, since one does not need sophisticated prior knowledge to work or experiment with it. He pressed the surface of the shape softly and said that they, for instance, looked at how various degrees of softness influenced the material's

---

14 I will go into more detail regarding the manufacturing of the RBO Hand in the following chapter, where I address the Hand's form.

behavior. The possibility of varying softness broadened the scope of possibilities, he explained. The silicone shape, which I held in my hand, was part of such explorations. The sub-optimal size relation of fingers and palm indicated that the shape was produced in order to explore how to direct and control silicone under air pressure and not yet how to accomplish grasping, which would require a different shape or additional material elements. The arrangement of these materials has been significant for the RBO Hand's development, insofar as it is a trace of the exploration of the characteristics of silicone.

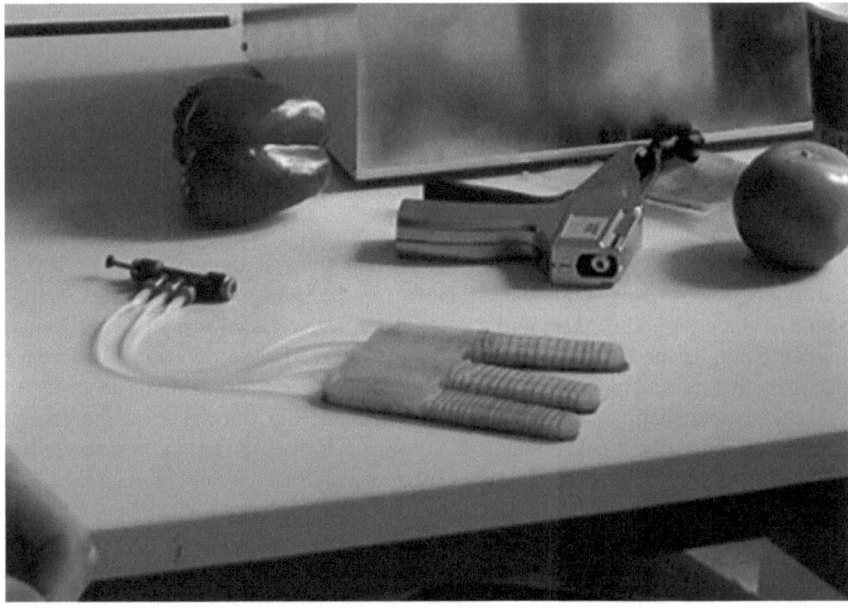

Figure 5: Hand-like shape in the laboratory's workshop (own picture).

As already indicated, the stories of ideas concerning the RBO Hand and *Mirage* share a similarity regarding their emphasis of deviation occurring in material practices. I outlined that there were material traces of such explorations in the laboratory workshop, where most of the manufacturing and tinkering takes place. In Baecker's studio, traces of past material explorations were ubiquitous.

"I might have a mechanical idea and then everything coalesces." (Baecker, *Mirage*)

2.1 Enacting the Past through Stories and Their Tangible Traces

Figure 6: Structure for testing the translation of the electric signal into a mechanical pull (own picture).

Similarly to the laboratory workshop, there were several general tools, such as a drill, screwdrivers, and different nippers. More specific to Baecker's practices were tools for working with electricity. These included, for instance, apparatuses like an oscilloscope and a sophisticated soldering station, as well as smaller tools like special nippers for stripping cables. Furthermore, there were several electronic parts stocked in the studio. These included small parts, such as cables and plugs, in addition to more complex ones, such as transformers and electric motor units. Alongside such analogue equipment, many parts were more specific to digital technologies, like a whole box of processors and Arduino boards.[15] In one corner of the studio, there were several wooden boards and metal plates, used shelves and iron bars. The stock of small bits like screws and hooks were too many to list. Most of that kind of equipment was sorted in labeled boxes or designated shelves.

There are no finished or already exhibited artworks stocked in Baecker's studio. He has a designated separate storage place for those. However, there are traces of his work in the studio that are more specific than the tools he uses. Prior to my first visit to the studio, I already knew some of Baecker's exhibited works. Hence, I recognized a pile of plates of acrylic glass, which were used for a previous installation. Baecker told me that they were cut for him and he uses the left-

---

15 Arduino boards have a large impact on Baecker's work. I elaborate on these in the upcoming chapter concerning *Mirage's* form.

overs as the basis for small prototypes and models. One of such models was arranged on his workbench (*Figure 6*). It was a plate of acrylic glass in which a hole approximately the size of a saucer was cut. There was a mesh of strings, comparable to a spider net, covering the hole. At the center of the mesh was a wire approximately one meter long and connected to a circuit board.[16] The circuit board produced a signal that made the wire contract and deform the mesh of strings accordingly. Baecker told me that he was using these leftovers to test how mechanical pulls translate between materials, such as between the wire and the mesh of strings. Without getting too precise, he went on and said that mechanical delays interest him, in the sense of signals propagating through different materials. This somewhat resembles his idea of giving image to the moving patterns of algorithmic learning. However, when I asked whether it was also a model for his new installation (which eventually became *Mirage*), he denied that it was a model or prototype; it was rather a test for a mechanical idea. He did not yet know if the idea would be sufficiently realized through the test setup.

At both sites, the laboratory and the studio, one can find material traces of inquiries that either relate to an idea for an object or that are significantly related to materializing ideas. However, only selected traces remain. Although it was probably only by accident that I ran into the hand-like silicone shape and the test setup in the studio, they shared a similarity that I consider significant to their remaining. That is, both were unfinished, but already embodied a technical character. They were assembled from different materials and arranged deterministically working technical relations. The hand-like shape deterministically bent, and the test set-up translated an electronic signal through different materials and actuated a mechanical pull. In this sense, both remained as technical units that indicated the material feasibility of an idea.

However, both units were not objects that spoke for themselves. They did not themselves produce a relation to past events, but required enactment to become relevant. Apparently, the spatial proximity of both units within working environments structured part of their enactment and made them easily identifiable as traces of events that took place there. Furthermore, it was Deimel's and Baecker's stories that made me realize the units' significance for inquiring material feasibilities. In that sense, both stories and materialities articulated what mattered and what did not. The stories enacted the basal technical configurations as matters of exploring ideas, not in the sense of finished technical artifacts that embody a concept, but as material traces of an idea's exploration. In this sense, stories of ideas and their material traces select differential patterns that unfold into past inquiries and articulate them as relevant to an object's becoming.

---

16   The wire is hard to identify in *Figure 6*. At the right bottom of the picture, one can see a small post. The wire is attached at its top and from there runs into the middle of the mesh.

2.2 Enacting Potentials through Figures, Prototypes, and Bodies

In the following section, I first leave the empirical sites of the laboratory and studio and turn toward published narratives, before I return to the locations to discuss the situated bodily enactment of ideas. The following sections share that ideas are enacted with an orientation toward the future, instead of an orientation toward the past as in the preceding section. Firstly, potentials are enacted through imaginaries and figures, then through a prototype, and finally through the bodily rendering of future objects.

2.2.1 Ideas and Their Figures

The narratives that I analyze in the following are texts taken from the project website related to the RBO Hand as well as a text written by Baecker for the exhibition flyer that accompanied *Mirage*'s first public appearance. Both texts were written prior to each object's technical realization. The website text concerning the RBO Hand was published on the institute's website right after work began on the project, whereas the text for the exhibition flyer was written a few months prior to the finalization of *Mirage*'s. What matters is not so much the date of publication, but that both texts enact the significance of both objects apart from their technical realization based on shared imaginaries and figures.

The Drowsy Human as an Ideal for Robotic Grasping

The institute's website[17] presents several research projects categorized by seven research domains. Among these, one is labeled "Compliant Manipulators," under which falls the RBO Hand. Besides pictures and web videos, the website also displays a short text that introduces the overall research objectives within the domain. The text starts with an episode that is not necessarily scientific, but is an imagined everyday scenario that is comprehendible to public audiences:[18]

---

17 This passage refers the structure of the RBO Laboratory's website: http://www.robotics.tu-berlin.de/menue/research/ (last accessed September 11, 2015).
18 The passage cited was copied from the website http://www.robotics.tu-berlin.de/menue/research/compliant_manipulators/ in December 2014. By September 11, 2015, the wording of the last sentence had changed slightly, and the website showed a picture of an advanced version of the RBO Hand, the RBO Hand 2 (cf. Deimel and Brock 2014).

"It is early morning and you just woke up. Sleepily you head over to your coffee machine, grab a mug and hit the switch. You don't waste a thought on what you just accomplished, while slurping down some black hot goodness.

Why is this seemingly easy process such an accomplishment? First, your delicate machinery of nerves, muscles and tendons changes its properties all the time, depending on whether you were just asleep, being alert, frightened, or just tired. At the same time, your senses might not give you reliable information about your environment, especially when being drowsy after just having woken up. Despite these complications, you grab your favorite coffee with ease!

We want to achieve the same grasping reliability, by using Compliant Manipulators."

The episode starts with a scenario that is probably comprehensible to all visitors to the website: making coffee while still half asleep. The protagonist of this imagined episode does not execute every single step of making the coffee consciously, but fulfills the task without thinking about it too much. The story interprets this mundane activity as an accomplishment. The text sees the accomplishment in the fact that we do not need to be aware of the complex workings of the human hand in order to fulfill easy tasks like making coffee. In such situations, the complex mechanics of the human hand work independently of our consciousness. Interestingly, this aspect of the story entails a paradoxical situation for the problem of grasping: on the one hand, grasping is a complex relation between hand, environment, and senses, and, on the other hand, human hands work somewhat autonomously in mundane situations. The ability of the human hand to cope with this problem with ease makes it a tool whose "reliability" is what the research institute wants to achieve "by using Compliant Manipulators."

I would like to point out two aspects of the story that relate the text to the larger context of robotic grasping: Firstly, the story has a protagonist whose senses are dizzy. This "drowsiness" is crucial to the analogy of the human protagonist to robotics research. In robotics, realizing specific dexterous grasping is less problematic than fulfilling easy tasks in undefined environments. This is reflected in robotics literature, which considers grasping in unstructured environments as a pivotal research issue of the field (cf. Balasubramanian and Santos 2014; Ben Amor et al. 2014; Dollar et al. 2014). In this regard, unstructured environments are comparable to the "drowsiness" described in the story, as it refers to malfunctioning sensors that cannot gather or proceed with sufficient information about the environment to fulfill a presumably easy task. In the text, this state does not seem to be problematic for humans, since mundane tasks do not require totally consciousness of every step taken. Secondly, the language used in the story is not arbitrary, but typical for robotics research. The anatomy of the body is described as "delicate machinery" that continuously changes its configuration in seamless and undetected interrelation with the environment and differing states of the body. The term "machinery" entails the complex workings of intricate mechanical elements. The precision and accuracy of the human hand makes it a unique tool that is admirable in terms of its delicate engineering. Describing the human body, and the human hand in particular, as an ideal "machine" is typical for literature on robotic grasping. In his book *Robot Evolution* Mark E. Rosheim has reported that the human hand's complexity has fascinated scientists and artists for centuries. He described the human hand in technological jargon as consisting of "a total of twenty degrees of freedom" and "driven" by approximately forty muscles (Rosheim 1994, 190). In that sense, the story figures the human body/hand as unique in terms of its complexity and autonomous functioning. The story uses language that addresses the reader directly and stresses that we tend to forget the admirable workings of our bodies, which enable us to fulfill complex tasks easily.

Both aspects, the problem of grasping in unstructured environments and the complexity of the human hand, are current research issues in robotics (cf. Bicchi 2000; Dollar et al. 2014; Ben Amor et al. 2014; Balasubramanian and Santos 2014). The story of the website provides a pictorial illustration of these problems. However, I think the story goes beyond illustrating problems in the following regards:

Firstly, the story positions the research idea in a larger robotics imaginary. The scenario reiterates the analogy of human and machine, which is a constitutive signifier of humanoid robotics and well described in science and technology studies (cf. Riskin 2003; Hayles 2005; Suchman 2007; Castañeda and Suchman 2014). The constitutive relevance of the figure "human" also holds true for the specific domain of robotic grasping, which is structured through continuous reiteration of what the crucial aspects of human grasping are in order to implement these in the design of robotic hands (cf. i.e. Balasubramanian and Santos 2014). Within the story of the website, the human-machine analogy converges in the term "compliance." It becomes a buzzword that signifies the approach of the RBO Laboratory in that domain. Surely, compliance belongs to the common terminology of robotic grasping (cf. Controzzi, Cipriani, and Carrozza 2014), but, in the case of the story analyzed here, compliance is signified by referring to an imagined human scenario, which is tropic, as it articulates associative meaning from diverse realms and categories of existence, namely between human everyday life and robotic grasping.

Secondly, the story unfolds the idea of solving the problem of robotic grasping from the same humanoid imaginary. It does not claim to copy human hands in general, which would also be a sensible thing to do in robotics, but proposes a focus on the relation of environment and the hand under conditions of insufficient sensory information. This focus stresses interactions between environment, hand, and sensors, and is made comprehendible through the drowsy protagonist, who wants to make coffee. In that sense, the story figures human grasping as an ideal, but equally does not propose mimicry of the mechanics of the human hand; it rather proposes a shift of the research focus, away from sensory information toward partial autonomy of the hand. In order to do so, the story does not draw upon the human hand as a category, but instead renders a human capacity. The story enacts the reliability of human grasping under the specific conditions of unreliable sensor information as an accomplishment that is admirable for robotics grasping. In that sense, the story is not only an illustration of research problems, but uses the imaginary human realm to signify the shift from grasping as a planning problem toward an interactional approach that builds upon a more competent hand. Hence, the story accompanying the RBO Hand is similar to the figure "human," as, for instance, described by Suchman. She has addressed how humanness is selectively enacted in robotics and AI research as categories of existence that signify the presumable essence of being human and, in this regard, the humanoid's boundary position between human and machine (Suchman 2007, 226ff.). Similarly, the story of the RBO Hand figures a specific human capacity as the relevant criteria for robotic grasping – human grasping becomes an ideal, in this sense. The story is tropic because it connects an imagined human scenario with the technical realm of robotics.

Signifying the Dream Story

When asked about the idea for *Mirage*, Baecker answered with the "dream story" I have already mentioned above. In that story, he indicated that the figures of his artwork are drawn from his inquiries into mechanical apparatuses and their combination with contemporary digital technologies. In contrast to the origin story of the RBO Hand, which Brock mainly characterized as a technological endeavor, Baecker did not separate his material inquiries from the figures he had in mind. For instance, he spoke of his intention to "build a machine that hallucinates." In his story, technologies and myth entail each other, which is indicated by the figural language he used to explain what he did at a practical level.

After the interview with Baecker, which was our first meeting, in May 2013, I wrote field notes about the story and was curious about how the figure of a hallucinating machine would change over the course of building the installation. Hence, I was surprised that the terms "dream" and "hallucinate" actually remained within the story that accompanied the finished installation. That story is a text authored by Baecker prior to *Mirage*'s first exhibition and was later published in a refined version on his website[19] and in the exhibition catalogue of *Mirage*'s first public showcase in April 2014:

---

19   Baecker's website: http://www.rlfbckr.org/work/mirage (last accessed September 11, 2015).

2.2 Enacting Potentials through Figures, Prototypes, and Bodies 87

"*Mirage* generates a synthesized landscape based on its perception through a fluxgate magnetometer (Förster Sonde). A fluxgate magnetometer registers the magnetic field of the Earth, which is dependent on the suns activity and feeds it into an unsupervised learning algorithm for analyzation. At the same time the algorithm that is based on the principle of a Helmholtz Machine 'hallucinates' variations of the previously analyzed signal. […]
I am speculating that the computers in the enormous Google data-centers cut off their perception (search queries, user behavior, speech recognition, image data) once a day and start to 'sleep.' What do their 'dreams' look like?"

In that story, the reference to the Helmholtz Machine, which Baecker had already mentioned in the interview in May 2013, remains. The reference is crucial because Geoffrey E. Hinton, creator of the Helmholtz Machine, made use of the wake-sleep metaphor (Hinton et al. 1995). Hinton applied the term to describe a class of artificial neural networks, which is a scientific domain that attempts to use biological neural networks as concepts for building algorithms. The discourse of artificial neural networks is prominently structured through anthropomorphic figures, similar to artificial intelligence as such (cf. Hayles 2005). Thus, the term "sleep," which Baecker used in the story that accompanied *Mirage*, is part of the same narrative realm applied in Hinton's texts concerning the Helmholtz Machine and AI in general. In this respect, one cannot state that the idea of building a "hallucinating machine" is something that has grown in Baecker's head alone, and he does not claim that this is the case. On the contrary, he even reiterates where he took that particular figure from and informs the reader about existing concepts that combine algorithms and the figures he uses to signify his artwork.

Still, the idea of *Mirage* seems original and individual in the text. I see the reason for this in the use of figurative language. Whereas the aforementioned story of the RBO Hand used an imagined scenario firstly to render a human capacity as an ideal for robotic grasping and secondly to embed this idea in a larger robotics narrative, the story that accompanies *Mirage* extends an existing narrative. The story picks up the wake-sleep metaphor of the Helmholtz Machine and goes beyond it by playfully leaving the reader with the question, "What do their 'dreams' look like?" In that way, the story embeds *Mirage* in the existing narrative realm of wake-sleep algorithms, but extends it by taking the figure more seriously than expected, as it moves from "sleep" toward "dream."

The choice to extend the narrative toward a machine that dreams or hallucinates is sensible for signifying an aesthetic installation. Unlike sleeping, both terms are related to pictures and images created through or within activity. Dreaming and hallucinating are both activities through which images evolve that are not controllable for the mind that produces and perceives them. Both terms implicate generating images without controlling them. This indicates an interesting tension for an image-generating installation. It indicates that the image, which is produced through the laser projection, is not a controlled artistic act, but a visual technical process that happens beyond human control and is caused by complex interactions among heterogonous agencies, such as the Earth's magnetism and digital algorithms. The text describes the visual image produced through such contingent technological processes as a "synthesized landscape." This trope is distinguishable and significant in its reference to a peculiar origin within the hidden life of the machine. In that sense, the trope "dream" connects *Mirage*'s visual aesthetics with a figurative account of a machine's hidden agencies.

The figural narratives of the RBO Hand and *Mirage* continue the object stories described in the previous section as well as embed these within larger technoscientific imaginaries. Through the figurative language, both ideas receive an orientation toward the future, which is concerned with potentials and not with origins. Furthermore, both stories individuate the objects. They reiterate culturally shared analogies between human and machine and use established narrative strategies. Nevertheless, both stories also break with the figures to which they refer. They select very specific aspects of shared imaginaries and alternate them. Through such narrative diffraction, both stories articulate shared technoscientific figures, past inquiries, and imagined potentials.

2.2.2 Embodying Material Potentials

> "There are many ideas and perspectives, but no dominant form." (Deimel, RBO Hand)

The robotics institute's website not only displays the text that I have interpreted above, but also displays web videos and pictures. Surprisingly, the website does not present a sophisticated version of the RBO Hand – not in the beginning of my ethnography in 2012 and not when I accessed the website in December 2014, at a time when the RBO Hand was already technically advanced. Rather, the website shows pictures and a web video of early attempts at using silicone as a material for grasping. One of these attempts was the starfish mold that Brock also mentioned as being one of the first shapes they produced after he bought the infrastructure to work with silicone. The web video is located shortly below the text and headlined with "Starfish Grabber."

The Starfish Grabber is made from a single piece of silicone consisting of six "fingers" (*Figure 7*). It is clearly far from being a sophisticated robotic hand – the Grabber is not part of a larger robotic structure, nor does it give the impression of being technically mature. Rather, it is left in provisional appearance – it is attached to a crude wooden stick, and its texture is left uncovered. Apart from its different shape, it is manufactured similarly to the hand-like silicone shape discussed above. It has an inflexible inner layer that makes the silicone bend inwards, and its fingers are wound with a thin thread to prevent them from blowing up. The web video shows the Starfish Grabber attached to the wooden stick and held by a human hand. The video starts by showing how the Grabber bends with air inflation. After fading in the headline "Starfish Grabber in Action," the human hand positions the Grabber close to an apple, which the Grabber easily captures with its silicone fingers. Then, another headline that reads "Grabbing from Suboptimal Positions" fades in. Now, the hand does not position the Grabber above but beside the apple. The Grabber captures the apple from this suboptimal position, too.

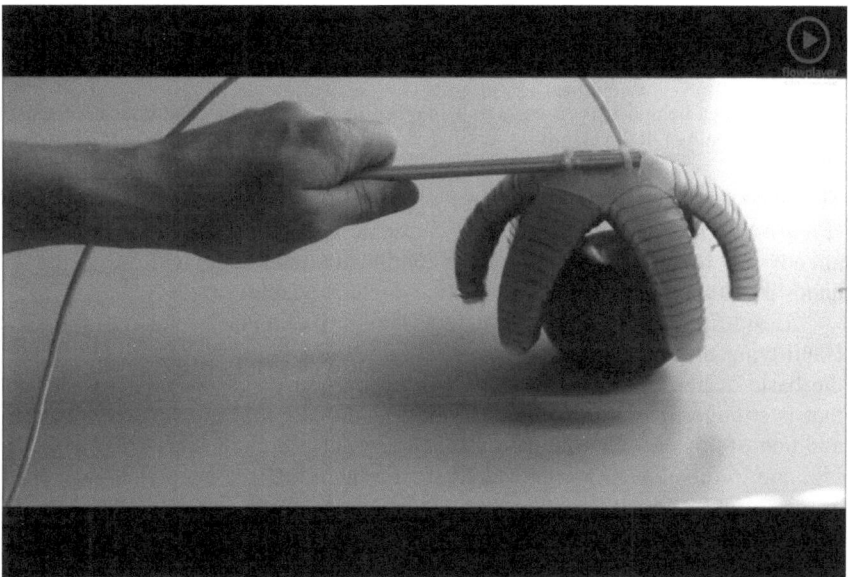

Figure 7: Starfish Grabber (source, Robotics and Biology Laboratory).

This last scene indicates that the Grabber's soft material is potentially beneficial for grasping under imperfect conditions, as the material easily complies with the new surface. In this sense, the Starfish Grabber somewhat materializes the human capacities figured in the story above. The Starfish Grabber's ability to grasp the apple from a suboptimal angle refers to the human capacity to grasp with insufficient sensory information. Clearly, the Starfish Grabber does not perform grasping comparable to that of a human hand, but the video presents what the text describes as an "accomplishment" that can – in principle – be realized for robotic grasping as well. Whereas the material trace of a hand-like silicone shape was enacted as a trace of exploring silicone, the Starfish Grabber is enacted through the video as an embodiment of the silicone's potential to perform reliable grasping based on interaction instead of sensory planning.

2.2.3 The Bodily Rendering of Future Objects

In the following, I show that the embodiment of ideas is not merely a storytelling practice nor exhausted by materializing a specific capacity. Rather, the enactment of ideas is a situated bodily practice. For this analysis, I draw on two video sequences recorded in the laboratory's workshop and studio (cf. Stubbe 2015). I first describe both situations before I interpret them together.[20]

> "The idea behind it is actually that interaction is of primary importance for grasping." (Deimel, RBO Hand)

The first sequence is an excerpt from an interview that I conducted with Deimel (*Figure 8*). The interview was part of the same visit to the laboratory that I have already reported above. The sequence continued after Deimel had shown me the hand-like shape. By now, we were sitting down at a table.

Besides its hand-like shape, there is another preliminary version of the RBO Hand lying on a metal box. This particular version is more advanced. It shares the basic design of the hand-like shape, but has an additional silicone element that is stronger than the fingers and molded like the ball of a thumb. With the addition of this element, the palm has a larger surface and better supports grasping, due to the element's rounded shape. Furthermore, the hand is connected to an air compressor and a computer so that its basic grasping function can be demonstrated.

---

20 The analysis of the video recordings has been methodologically informed, in particular, by Charles Goodwin's sequential interpretations (Goodwin 2000), Hubert Knoblauch's focused ethnography (Knoblauch 2001), and Lorenza Mondada's focus on multiple temporalities that conflate in material practice (Mondada 2012). Charles Goodwin's studies also point out ways to compare small-scale interactions. For instance, he compares interactional patterns in sequences of young girls playing hopscotch with archaeologists classifying colour.

2.2 Enacting Potentials through Figures, Prototypes, and Bodies 91

Time	Still of the Video	Transcript
I. 00:55		Deimel: "*You only need one signal: inflating, releasing. But you can make very complex deformations from that. This is usually not done in robotics. Typically, electric motors have very good, linear characteristics. With these rubbers, many interactive things happen with the environment.* [presses some keys; the silicone hand coils up]
II. 01:45		[positions his spread fingers between the finger tips and palm of the silicone hand] *It is soft.* [puts an artificial apple into the silicone hand] *When something gets into the hand, its form adapts to it.*
III. 01:56		[mimics a round form with his hand; the apple rolls out of the silicone hand] *This is exactly what we want to make use of here. That the hand is not steered; where the fingers have to be or how much pressure or power has to be applied. We create another kind of communication of the hand.*
IV. 02:24		[takes the apple in his hand, waves it, and puts it back onto the box] *And we just try to establish as much surface for contact as possible. The more contact surface you have, the better it grasps. Surely, it does not always work, but this is generally the basic principle."*

Figure 8: Enacting the silicone's softness and potential for grasping (own video).

Deimel opens the conversation by explaining the basic advantage of a robotic hand made out of silicone, which he sees in the low signal processing needed to enable complex grasping. He points out that this is not a typical approach for robotics, which is still dominated by hands operated with electric motors. A crucial difference between his Hand and others is the "many interactive things" that happen in relation to the environment. He enacts this in Frame II, in which he demonstrates the Hand's softness by easily spreading its fingers with his fingertips and placing an artificial apple into it in order to show its ability to adapt to its environment. After this practical task, he continues in Frame III by explaining the more abstract principle behind it. His and his colleagues' idea is to "create another kind of communication" for a robotic hand, as opposed to steering it. In the last frame, the pragmatics of this "communication" are again emphasized by referring to the importance of the contact surface for good grasping. The fact that this "does not always work" highlights the exploratory character of their novel approach to grasping. The sequence is a bodily enactment of the silicone's softness and a demonstration of the material's potentials for robotic grasping. Deimel's body is an interactional resource that connects the material artifacts at hand, the conceptual ideas behind the hand, and what he regards as my expectations regarding robotic hands in general.

"The idea now is that you place an elastic element somewhere here." (Baecker, *Mirage*)

The second sequence shows Baecker and me in his studio (*Figure 9*). We are not doing an interview as I did with Deimel, but casually chat while I observe him during his mundane creative practice. Baecker is working on a new test structure. This new test structure consists of different materials than that which I have addressed as a material trace. It is made from two wooden plates, hooks, threaded bars, strings, nuts, and bolts, as well as customized pulleys and an elastic element. The two plates are attached through the threaded bars; the space between them is approximately 40 cm wide. On each plate is a grid of hooks onto which the customized pulleys are attached. One string is run through several pulleys from one plate to the other. At one end, the string is not directly tied to a pulley but to the elastic element, which is connected to a hook.

The sequence begins with Baecker explaining how the new test structure works. First, he points to the elements that are already in place. He signifies these by referring to "a kind of delay" that is supposed to evolve through a specific ordering of strings, pulleys, and the elastic element. In the second frame, he starts to refer to the anticipated aesthetics, which are supposed to evolve through a kind of movement that appears to propagate through the structure. By moving his arm like a snake, he mimics what kind of behavior he would like to achieve. He emphasizes that this is a challenge. In the following frame, he relies on my ability to imagine what he has in mind. He expects me to imagine how he will continue to build the test structure in order to figure out if he will be able to

2.2 Enacting Potentials through Figures, Prototypes, and Bodies

Time	Still of the video	Transcript
I. 02:35		Baecker: *"When you attach this one, then it moves...* [points to the pulley] *The idea is now, if you build in an elastic element; so when I make the movement here, it arrives over there three seconds later, as a kind of delay. Then a string is tightened between the parts.*
II. 03:06		*Then you have a kind of line, which propagates through it.* [makes a snake movement with his hand] *And to have that in several dimensions, so you have a row. I would additionally hang that separate.* [orientates his gaze and hands towards the wooden panels] *... so this is hard.*
III. 03:35		*Imagine this was there in every row, like four, five times, then I would replace this hanger with that.* [makes a bow movement with his hand] *So the whole system is hung in two dimensions, totally detached, totally sprung. Only at every entrance does a signal enter.* [points to the end of the wooden structure] *So there is a motor or a cord entered here and here.*
IV. 04:10		*And you actually have a wafting area. You have a wafting area through which this is wandering through slowly. The best is a closed circuit that is only triggered once. I always had something like a landscape situation in mind."* [continues to pull the string through the eyelets]

Figure 9: Enacting the "landscape situation." (own video)

establish the propagating movement within a closed circuit (Frame III). Furthermore, I should also grasp the image that he has in mind. The first image is a "wafting area" – an image close to the movement of strings, which lie partly assembled in front of us. The second image is "a landscape situation" that he had in mind as an initial idea for the installation. In the sequence, Baecker's body connects the rudimentary test structure at hand with the figures of his artwork. By embodying and acting out the "landscape situation," the imagined image becomes part of the shared reality between him and me.

Both situations stress consideration of the objects' enactments as framed through interaction. In particular, the stills of the video recordings capture distinct bodily activities in these interactional framings. Firstly, there are pointing gestures. They accompany explanations and indicate what Deimel and Baecker are referring to when they talk (cf. Goodwin 2000). In the second sequence, for instance, these pointing gestures select those parts of the structure that are described as crucial to establishing the anticipated movement. Secondly, there are gestures that mimic and physically enact the future object. This bodily simulation is a distinctive form of enacting epistemic objects that have not yet materialized, but that are referenced in communicative situations. Myers calls such bodywork "embodied imagination" (Myers 2008, 165). In her study on protein modeling, she argues that material and mental models are not to be regarded as dualistic, but rather as deeply entwined. Through embodied imagination, researchers incorporate the inner structures of models and enact them as epistemic objects. They use their bodies to make graphical objects tangible, and they employ gestures and movements in communication with novices in order to flesh out and relay their knowledge about otherwise only virtual objects (Myers 2008, 180). In both sequences, I am such a novice. In the first sequence, Deimel uses his body not precisely to mimic the Hand, but to enact the distinctive difference of his silicone Hand from how he expects me to think robot hands typically work. In order to do this, he demonstrates the softness of the silicone Hand by easily spreading its fingers with his. He does so without force or additional programming, so I can comprehend the Hand's compliance. As already mentioned above, this compliance is a basic principle of the Hand's distinctive kind of grasping. Nevertheless, here it is not referenced through a story like above, but through Deimel's distinctive bodily movements. In a similar but not identical way, Baecker makes use of his body, not to enact the test structure as such, but the idea behind it. The snake movement in the first sequence (Frame II) gives body to the not-yet-realized aesthetics of the future installation. Its realization might still be far away, but the image of "a kind of line that propagates through it" already structures the situation at hand. Both situations are co-produced by my bodily presence, which is reflected in the specific gestures Deimel and Baecker use to enact their objects. I cannot tell whether they would have acted similarly toward somebody else (probably yes), but, crucially, what the sequences indicate is that articulating an idea is a situated practice across material assemblages, bodily movements, and accompanying stories.

2.3 Articulating an Object Identity

For the preceding empirical analysis, I regard ideas as articulations. This stresses what ideas connect instead of what they make vanish. This is how ideas connect heterogeneous elements from different temporal, discursive, and material realms to create a temporal sense of unity. That perspective is the opposite of regarding novelty as invention, which would consider ideas as plans or objectives. Furthermore, the perspective pushes ideas back into the scope of science and technology studies, because articulations begin by treating ideas as actors' categories that *do* something. In the following section, I approach the question of *what ideas do* in more abstract terms than before in order to delineate how novelty becomes part of a shared reality. In the following, I sum up four typical articulations of ideas from the preceding section before I turn to the work they do in the remainder of this paper. I regard this work as building an object identity.

2.3.1 Four Articulations of Ideas

First, an idea articulates *selected pasts*. The stories in the first section enact past events, inquiries, and explorations as relevant to how the idea for an object emerged. They do not enact any elements, but only those that mattered. What mattered to both actors was the deviation that occurred in material practices. Brock emphasized that ideas developed over a longer period, through different experiments and intersections. Baecker similarly reported that a finished object was always different from an initial idea, and this deviation made him investigate something further. In this sense, an idea is not a narrative element that erases the trajectory of an object, as Latour and Woolgar would stress. Rather, ideas mark selected pasts as relevant and make differential patterns accountable for an object's becoming. This concerns the general production of temporality. The narrated histories of the RBO Hand's and *Mirage*'s origins produce their own temporal order by marking what matters and what does not. This marking refers to an idea and relates its formation to past events. By visiting places of inquiry, one encounters that marking is not only a narrative practice, but also entails material traces that sediment past inquiries.

Secondly, ideas are figurative, *which makes future objects relate to and contest shared imaginaries*. This is shown in both ideas' enactments as part of larger technoscientific imaginaries. In the case of *Mirage*, the imaginary was the story of the idea's initial expression. Baecker mentioned the dream story and a machine that hallucinates as the idea that drove him to build *Mirage*. In the case of the RBO Hand, imaginary and idea seem to have a different temporal relation. Brock regards the idea as a logical consequence of inquiries, and its positioning within a larger imaginary is more a second-order storytelling that signifies the research approach. However, despite these differences, using tropes made both future objects coherent, as they relate their respective ideas to a shared imaginary that signifies the object's potentials. This is not to say that ideas were mimetic or representational. Rather, tropes enact ideas as capacities that signify an object's future existence and, in particular, its difference. Figures not only relate anthropomorphically to technical realms, but also extend, modify, and contest them in order to render specific human capacities significantly different from the common interpretation of that imaginary.

Thirdly, an idea's potential is not only a narrative promise, but is *co-produced through concretization*. This meaning of ideas requires movement beyond an idea as a semantic term and stresses consideration of material processes that open an object to meaning-making. The pragmatic entailment of idea and materialization is shown, for example, in how Baecker reports that he "might have a mechanical idea and then everything coalesces." In this sense, an idea and its materialization have a twofold relation: a) ideas are inscribed in the engineering of an object, and b) technical elements need to be assembled in terms of their internal resonance in order to open up for attribution (Simondon [1958] 2012a). The second relation was the condition for the Starfish Grabber to exist as a meaningful object. Its embodiment does not resemble a robotic or human hand, but, foremost simply works as a functioning unit, since the unit has the capacity to grasp an apple from a suboptimal position. This technical concretization opens the object to articulating the feasibility of compliant manipulators and its potential for advancing robotic grasping. In this sense, concretization co-produces an idea's potential.

Fourthly, ideas are *situated enactments*. The bodily rendering of ideas brings into focus what the previous sections have implicitly carried or marginally noted: ideas are enacted in specific situations in response to expectations and immediate interactional resources. Both sequences capture how expectations co-produce the performance of what is original or different. I, a researcher who is interested in the actors' projects, am inscribed into what is regarded as interesting or worth emphasizing. Both actors address me personally, which entails addressing my expectations. This includes gestures that mimic or perform objects and respond to my bodily presence. Furthermore, in both situations, future objects are enacted through several interactional resources as stories, materialities, and bodies. Such resources give body to an idea and connect what was once separated. Here, ideas articulate immediate situations of materialities, expectations, and bodies with the images and potentials of future objects.

2.3.2 Object Identities

Now, I want to turn to the work ideas perform and reflect upon the tension between individuating and relating objects. Whereas the preceding four points make up a summary of the actors' meanings of ideas, now the focus is on what these articulations do to the object in terms of its novelty. In the following, I approach this question through an analogy.[21] This analogy is Mead's theory of the human self (Mead [1934] 1967, 135ff.), which I use to think through the ways in which an object becomes something coherent *and* significant. My claim is that ideas, as I have rendered them in the preceding section, work toward an *object identity*.[22]

The Generalized Other: Biographical Trajectories and Kin Objects

Mead's theory is one that is concerned with the genesis of the self as a social process. He began his theory by sketching the initial structure from which this genesis proceeds. In this step, his theory is already marked as one that focuses on

---

21 Using an analogy is a way to learn something about what is not yet understood through something that is better understood. In this sense, I want to learn something about objects and novelty, and, in order to do so, use Mead's theory of the self to find interesting similarities, connections, or differences in comparison to my analysis. This methodological approach is inspired by Strathern's style of analogical comparison (Strathern 1991; Strathern 1999; Morita 2014). To clarify, I do not transfer Mead's theory onto my object theory, which would entail claiming that all *a priori* assumptions (i.e. that it would be nonsense to assume the relevance of thought processes for objects), as well as entailed processes, match. Instead, I use selected elements of Mead's theory and ask if I can find similar patterns or movements in my material that are significant to an object's novelty in a state of previously disconnected elements.

22 Ideas are part of Mead's theory, too. Mead's understanding of ideas as responses to social demand (Mead [1934] 1967: 180/1) is not contradictory to my analysis. Nevertheless, I see no epistemic potential in transferring his understanding of ideas into my analysis, as this would entail giving up the inductively articulated four meanings of ideas. Mead's understanding of ideas is too abstract and restricted to his theory of human conduct; the primer offers no epistemic resistance for my analysis, and the other would transfer my focus onto conduct and away from articulations. Hence, when I speak of ideas in the remainder of this study, I mean my account and not Mead's.

the social as explicans. He stressed that what people consider a self is constituted by the collective's and individual's relation to it. Crucial to relating to the collective is language and the use of symbols that stimulate response. Through reflective observation of such responses, and the ability to take on another's role, one can find or build his/her relation to the collective. A children's play already entails this process and Mead considers play an initial phase in the genesis of the self. Whereas play already entails self-awareness, it is the game that requires handling of the attitudes of the some toward another. In a game, one needs to organize the multiple attitudes of different players in order to participate and compete. In order to cope with this complexity, we act upon the collective in a generalized way – upon the "generalized other," in Mead's words. The structure from which the self is built is the response of the generalized other. From this understanding, being part of a community is the initial condition for the self's becoming, and the self is immanently social (Mead [1934] 1967, 164):

> "No hard-and-fast line can be drawn between our own selves and the selves of others, since our own selves exist and enter as such into our experience only in so far as the selves of others exist and enter as such into our experience also."

What does this mean for the work done by ideas? Mead stressed that language is a means through which one relates to another. In the beginning of this chapter, I stress that an idea is foremost a category in the actors' language. In their first articulation, ideas structure events as they unfold the trajectories of the inquiries and experiments that have proliferated something deviant. By referring to an idea, the not-yet-realized RBO Hand and *Mirage* are placed in relation to selected past events and receive a biographical origin. This is not a causal relationship and does not determine the path a technology takes. On the contrary, telling origin stories is a matter of constructing coherence or a sense of unity in an otherwise messy past – and language is a means to doing so. Furthermore, language not only structures the biographical route an object has taken, but also a future object's position within a generalized collective of pre-existing objects. For instance, the story of the RBO Hand's origin continuously refers to the Hand as resulting from a research process within the field of robotics; it refers to the problem of grasping in unstructured environments, which is a shared problem in the robotics community and to which the Hand is a response. Similarly, the idea of *Mirage* not only signifies the material inquiries that took place in an isolated studio; additionally, inquiries led to attempts at building objects that were similar to scientific apparatuses, and hence referenced technologies that were not akin to *Mirage* by origin, but via the desired behavior. As such, the idea for *Mirage* articulates the Helmholtz Machine into a figurative story. In this sense, the ways ideas work upon objects is analogous to the power of the language and shared symbols that constitute the genesis of a self: they articulate an object's biographical trajectory and the generalized collective of kin objects, both of which are the initial conditions for an object's identity. In this sense, ideas organize the generalized others that exist and signify the game that is played.

Becoming Distinct through Rituals

The self, or an object identity, is built upon this background. Mead emphasized that the self is built from interaction among individuals, which in return entails that the collective exists before the self. Nevertheless, having said that, the question remains as to how something becomes distinct. Addressing this question, Mead continued his argument by stressing that the self arises when one takes over the responses of the others into one's own conduct. When the attitude of the others affects conduct, in the sense that one takes on attitudes and replies to them with corresponding gestures, then a subjective self comes into being (Mead [1934] 1967, 167). This is a matter of becoming distinct from others and taking on attitudes requires self-consciousness. In that sense, Mead said, being self-conscious and reflectively responding to one's own position in conduct means becoming an object to oneself (Mead [1934] 1967, 172).

However, searching for an analogy between self-consciousness and the genesis of an object identity has little potential for understanding how an object becomes distinct, as Mead's theory draws on a thinking and reflective human being that is very unlike the objects that I investigate here. Despite this undisputable difference, there is a passage in Mead's theory in which he proceeds from the question of how one becomes subjectively distinct toward the special case of how one reacts against the disapproval of the collective. For Mead, there is only one way in which we may change the attitudes of a whole group, which is "setting up a higher sort of community which in a certain sense out-votes the one we find" (Mead [1934] 1967, 167-8). In so doing, one can stand out as oppositional and go against the world with the potential to change it, or at least the other's attitudes. Integral to convincing the other is speaking with a voice of reason, with "voices of the past and of the future" (Mead [1934] 1967, 168). For Mead, this is the only way the self can obtain a voice that is more than the voice of the collective. With a voice of reason, one is not simply bound by the collective, but may reform the order of things. This speaking out is embedded in social rituals. As an example, he drew upon a day in court, which is a critical situation wherein the self is oppositional to the judging other. The defendant may present his/her views in order to change the other's attitude. Speaking out is not only a right one has, but foremost a duty in order to legitimately change the attitudes of a community. The self is mutually oppositional and part of the collective in this situation.

Bringing about changes takes place through the ritual interaction that enacts the distinct positions and roles of participants.

This episode in Mead's theory on how the self becomes subjective has more potential as an analogy for an object identity than the previous remarks on self-consciousness. By comparing Mead's theory with the four articulations of ideas, which I outline above, one can draw connections between how ideas organize the others that exist and how they make an object distinct through basic rituals before it is fully realized. Similarly to Mead, the first and second articulations focus on the use of language. In the first articulation, ideas relate selected past events and future activities. The crucial situation of their enactment is the basic methodological procedures that have brought such stories about: the interview. The interview requires the ordering of past events and signifying the object that is about to become. In light of Mead's remarks, such situations are rituals in which selected past events are presented as stories of reason that legitimate the position of the self. In rituals such as an interview, one has to mark what is distinct about a new object and how its characteristics are a matter of directed explorations. The object becomes part of a biographical trajectory and distinctly different from what came before. In their second articulation, ideas are stories that build upon figurative accounts of shared imaginaries. Mead regarded such stories as "setting up a higher sort of community." These stories embed a future object in shared imaginaries, but also go beyond them, as they emphasize disregarded elements or extend a narrative. Both strategies entail abstraction in order to set up a higher form of narrative that mutually allows embedding an object in a shared imaginary and alternating the imaginary. This strategy is similar to the defendant in court, who gives an account of law through abstracting and relating elements of his own crime. In this sense, the figures of the RBO Hand and *Mirage*'s accompanying stories are legitimately oppositional, as their deviance builds upon the beliefs of shared imaginaries. Surely, the analogy is limited so far as ideas do not make an object fit the common standard like the defendant would try to in court. What the analogy allows is the connection of ideas and the ways an object is justified as meaningfully different. In this sense, ideas articulate an object identity, as they embed an object in the imaginaries of a collective and mutually give reason to its opposition.

"I" and "Me" as Meaningful Potentials

Continuing with his theory, Mead elaborated on the mutual character of being part of a collective and becoming distinct. He captures the organization of both with his prominent distinction between the "I" and the "me." Whereas the "I" reacts to the self, the "me" is the organized set of the other's attitudes to which

2.3 Articulating an Object Identity

the "I" reacts (Mead [1934] 1967, 174-5). Mead considers the "I" as the part of the self that responds to attitudes it is confronted with – it is impulsive compared to the "me," which mirrors the attitudes of the collective: "the 'I' gives the sense of freedom, of initiative" (Mead [1934] 1967, 177). In that sense, the "I" is the part of the self that causes diffraction as it moves into the future and pushes off expectations. Mead wrote that the steps of the "I" are "in a certain sense novel" (Mead [1934] 1967, 177). Whereas the defendant in the example above mainly attempts to find a language that legitimizes his opposition, the "I" is pre-social and not concerned with fitting into the given order. On the contrary, fitting expectations is how the "me" structures the self.

In the object stories of the first articulation of ideas, I addressed deviation as a central narrative element. Whereas I consider deviation as a narrative construction above, in the sense of giving reason to the biographical trajectory of an object, Mead's theory of the "I" suggests taking the material inquiries of the object stories for granted. If I take the stories of deviation for granted, the inquiries Brock and Baecker have reported on were structured by an idea with a strong "I." Brock reported that the idea of the RBO Hand was a consequence of experiments concerning a different topic. Their experiments enacted resistance that was accommodated by following a new direction (cf. Pickering 1995). Similarly, Baecker's inquiries responded to deviation between his idea and the contingencies of material practice. Analogously to Mead's "I," these material inquiries entail resistance that enforces movement and changing of the given order.

However, this is one way to use the analogy of the "I" and "me" for an object identity. However, I do not want to go deeper in that direction, as it entails giving up a critical stance. Still, there is another aspect in Mead's theory of the "I" and "me" that is significant for the work performed by ideas, as I understand them here. This aspect is the relation between impulses through material inquiries and the expectations of fitting a specific object type or label such as "robotic hand" or "media installation." I have already outlined that ideas organize the others that exist and how they embed an object into the imaginaries of a collective. Whereas this aspect focuses on how ideas become part of shared meanings and manipulate them through language, the analogy for the "I" shifts the focus onto micro scales of accommodating ideas. This accommodation is addressed in the third meaning of ideas, which captures the co-production of ideas through material concretization. The example of the Starfish Grabber shows how an idea gains potential through material assemblages with a technical character. The Grabber's functionality indicates that an idea has the potential to solve a collective problem – namely, solving the collective problem of grasping from suboptimal positions by using silicone as a material for robotic grasping. This collective problem is similar to what Mead regards as the "me," because the problem articulates expectations of a new robotic hand. The potential of the silicone responds to these expectations, not in the sense of a defendant as referred to above, but by enforcing a new material form, the Starfish, that is uncommon to robotic grabbers but responds to the collective problem. In this sense, there is a reciprocal relation between material impulses and collective expectations. The work done by ideas is analogous to coping with the struggle between the "I" and "me." Ideas articulate impulses and expectations as they enact material potentials in a form that makes them meaningful to the collective.

Embodiment and Materiality

These analogies raise questions regarding what role materiality and embodiment play in building an object identity. So far, I have focused on the power of language and only stress in the last paragraph how ideas articulate material potentials and expectations. Left out is the embodiment of ideas, either through material objects like the Starfish Grabber or bodily enactments as captured in the fourth meaning of ideas. Surprisingly, the role of the body is somewhat neglected in Mead's theory of the self. Although Mead indicated that gestures and organisms are implicated in an individual's response to the world, he conceptualizes the genesis of the self as a cognitive process (cf. Gugutzer 2001, 70). This is surprising, so far as Mead was a social theorist who largely attended the physicality and materiality of social processes. If I change my method for now and do not treat Mead's theory as an analogy between self and object identity, but rather take his theory literally as a social theory of conduct, then I can use his remarks on symbolic interaction to better understand how Deimel and Baecker enacted their objects bodily and how this enactment pushed forward an object identity. For this interpretation, I take up how Mead addressed materiality in his philosophy of conduct (Mead 1987).

Mead regarded the physical environment as not exterior to the mind; rather, he regarded our bodily response as always implicated in how we act toward material things (Mead 1987, 88ff.). For Mead, we identify the universal character of things by anticipating how we respond to them. Our response is not naïve, but meaningful through experience and the significant gestures and symbols that we use to communicate a thing's character. In that sense, our bodily response to things is included in how we communicate an object's meaning (Mead 1987, 103). To a certain extent, there is not much new coming to the forefront if I retell my account of the bodily rendering of ideas with Mead's words. I have already addressed how bodily gestures respond to Deimel's, Baecker's, and my bodily presence, and furthermore how they used their bodies to either act out the

2.3 Articulating an Object Identity 103

physical character of an object or to enact an object's material behavior. In both situations, bodies communicate an object's character through meaningful gestures that consider bodily presence, just as Mead said.

However, there is another aspect of Mead's theory that he might not have focused on explicitly, but which I would like to push to the forefront. It is how the object's technical character transduces[23] diverse realms (cf. Simondon 2009, 11). Mead randomly addressed this aspect by mentioning how our anticipated physical response is continued in conduct (Mead 1987, 95). By transducing, I mean how the silicone's behavior and respectively the behavior of *Mirage*'s test structure continued in Deimel and Baecker's bodily movements. Though the bodily rendering of ideas, Deimel and Baecker flesh out the object's physical behavior that is not yet realized, but which becomes part of a shared situation through its bodily enactment. Deimel enacts the silicone's potential for robotic grasping by using his fingers to demonstrate its compliance. Similarly, Baecker continues the movement of his test structure by mimicking a snake movement with his arm. In both situations, gestures act out those parts of an object's technical physicality that are not yet fully realized, but that are meaningful to an object's novelty, as they signify its potential to be different. These unrealized characteristics of an object become part of a shared reality by continuing the partly realized behavior of material elements in another realm, the body. In that sense, Deimel and Baecker respond to an anticipated physicality that implies their experience of physically engaging with the materialities at hand. Their bodily movements push forward an object identity as far as they enact the missing, but meaningful, physicality of an object.

To sum up, analogously to and in dialogue with Mead's theory of the self, I delineate four aspects of an object identity articulated though ideas. Firstly, ideas organize the others that exist, as they signify an object's biographical trajectory and reference its generalized collective of kin objects. Secondly, ideas articulate stories in rituals, which embed an object in the imaginaries of a collective and mutually give reason to their opposition.[24] Thirdly, the work performed by ideas is the articulation of impulses and expectations, as they enact material potentials in a form that makes them meaningful for the collective. Fourthly, ideas respond to situations, as they transduce different material realms. In all four of those aspects, ideas articulate an object identity as they simultaneously individuate and relate an object; they are necessary for articulating coherence in the diversity of materialities, stories, and bodies that make an object.

---

23 Transduction is an operation that propagates from one element to the next (Simondon 2009: 11).
24 In the third articulation of novelty, I address biographical rituals, too. There, I address them as passage points, like *Mirage*'s exhibition. In contrast to the identity building here, later passage points require an object's realized efficacy.

2.3.3 Novelty as Object Identity

I open this chapter by stressing that ideas have somehow vanished from the rhetoric of science and technology studies. Either they are neglected as belonging to the perspective of novelty as invention, and hence lie outside the ethnographic interest in differential patterns, or they are shipped around as they evoke associations of paradigms that are not concerned with the "dirty" practices of laboratories. In the preceding section, I show that ideas can be beneficial for investigating the becoming of technological objects; given, one takes them seriously as what they are in the first place: an actor's category – a trope that belongs to the stories of an object's becoming. In that sense, the analysis of ideas is a methodological access point to studying how symbolic, material, and bodily elements articulate a sense of unity in an otherwise messy constellation. Furthermore, using ideas as a starting point creates comparability, as they make up an empirical category that appears in various situations across cases and sites. Such an approach to ideas is not contradictory to the practice orientation found in science and technology studies, nor in the more general perspective of novelty as differential pattern, because it does not take ideas for cognitive plans or intentions. Rather, it builds upon the symmetrical perspective that signifies the methodology of laboratory studies and takes into account the diverse realms via which an object exists.

The genesis of an object identity takes conceptual inspiration from the perspective of novelty as biographical passage. Object identities, as understood here, share with that perspective a concern with novelty as a process of meaning-making and the conceptual analogy for human socialization. In contrast, I focus on activities like selecting and re-arranging diverse elements so as to create meaning that takes the response of generalized collectives into account. This is different from the approaches summarized as biographical passages so far, as it connects accounts of collective meaning with the future orientation of actors and their prototypes. In this regard, object identities can be related to the discussion of scenarios and expectations in technological development (cf. Lente and Rip 1998; Lente 2012; Schulz-Schaeffer 2013). This discussion elaborates on how imaginaries and scenarios coordinate collaborative actions. My analysis of ideas relates to this discussion as far as it shares the interest in stories and how ideas build a reality that becomes the symbolic habitat of new technologies. Whereas the discussion mainly remains on a macro level, I add that such realities not only repeat on a macro scale, but are foremost negotiated with individual objects and experienced in immediate micro-scale interactions. The object identities delineated here do not simply relate to a larger scenario or general expectations, but always contest these. Such contestation is an interactional practice across diverse realms and signifies an object's novelty.

Novelty, in this regard, is the articulation of meaning. It goes beyond language, since an object's individuality and difference is *made* with materialities, which remain as traces of differential patterns or embody and signify potentials. *Novelty as object identity is the* sense of unity *that connects biographical trajectories, shared imaginaries, and generalized collectives, as well as materialities and bodies that come together in interactional situations.* It is a shift from the unconnected toward joined elements with a shared identity.

3 Form: How Materials Become Effective

"Ce n'est pas avec des idées qu'on fait des vers, c'est avec des mots."
Stéphane Mallarmé (1842-1898)

It seems as though the French poet Stéphane Mallarmé challenged what I outline in the previous section. "We do not make poems with ideas, but with words," he wrote. Mallarmé's poetry is concerned with the force of rhythm, sound, and changing meaning depending on if a word is read in silence or aloud. In contrast, I have followed ideas that articulate an object through diverse realms and delineate how something coherent, something with an identity, comes into being. Mallarmé's approach to poetry was revolutionary, as he took the opposite approach, but nevertheless with a similar aim. He crafted language and put words together according to their tonal and musical qualities, not focusing on their shared meaning in the first place. Playfully, he conducted the ambiguous meanings this new grammar produced; he described this inventive approach to language as "paint, not the thing, but the effect it produces."

Whereas linguistic sciences regard Mallarmé as the most symbolic of symbolist writers, I would regard his approach as technical – and not exactly technical in an instrumentalist sense, as a means to an end, but in Simondon's sense as guided by an internal fitting of elements. The initial concern of such an approach is less about instrumental application, and more about growing the organicity of single components that form a functional unit according to their internal resonance – like a chain, sprocket, and hub that form a drivetrain independent of its application in a bicycle or motorcycle. By assembling elements according to characteristics that enfold in their coupling with others, objects gain a technical character, or *technicity* (Simondon [1958] 2012a). However, growing technicity does not close a technical design in its shared meaning; on the contrary, it opens an object up to external information and attribution. As in Mallarmé's approach to poetry, technologies become meaningful through the relation of their elements and their efficacy as a novel unit.

In the following section, I focus on how such technical units evolve. Similarly to the preceding chapter, embodiments play a crucial role – however, not as enactments of ideas and potentials, but as technical forms that realize a working of material elements together. The focus is how technical elements and features are assembled according to their fitting, and how they become technology in the sense of concretized relations in a functioning unit. In the previous chapter, objects are articulated as ideas that made previously loose relations coherent through diverse realms. Retrospective stories, material traces, discursive figurations, and bodily renderings structure the becoming of an object identity; these diverse enactments share that they are able to enact an object according to its future potentials, in the sense of giving meaning to what is about to become. In contrast, the present section focuses exactly on this becoming. However, this is not to be mistaken as a second stage or subsequent step. Rather, it articulates novelty in a different form – not as signifying the difference between what is given and what is potential, but as a technical form whose novelty is articulated through its efficacy. This could also occur in early prototypes of an object or through the advancement of already applied technologies. What makes such articulations distinct is the technical character of their enactment and their physical functioning.

What do I mean by stressing a technical form? Speaking of form seems a little bit old-school. Form brings to mind the four causes of Aristotelian thinking. According to these, change and movement result from a) matter or materiality, such as, for example, the gold in a ring (*causa materialis*), b) the form or shape a material enters, such as thin, round, and hollow (*causa formalis*), c) the end, as symbolizing marriage (*causa finalis*), and d) the effecting cause, which brought about a change or movement, such as the jeweler who crafted the ring (*causa efficiens*). Stressing the technological form appears as singling out one of these causes and limiting explanations to one determining element. Indeed, in this section, I focus on how an object enters a technical and material shape. Singling out one cause in order to ask what technology is has been widely criticized. Heidegger, for example, stressed that the causes blind what connects them. Singular causalities block asking for the technological "essence," which he regarded as a "revealing" of the world (Heidegger 1977, 6ff.). Inspired by Heidegger, but from a sociological perspective, Rammert has urged that one cause always falls short of explaining what constitutes technology. Only taking into account material, technical, functional, and practical dimensions may capture the realms through which technologies exist (Rammert 1998, 294-5). It is not my aim in this chapter to contradict these opinions; on the contrary, I agree with and even build upon some of their issues. However, I see questions of form as not sufficiently addressed, particularly when inquiring into novelty and innovation. Social science perspectives eagerly stress how material resistance causes differential patterns (*causa materialis*) (i.e. Pickering 1995; Rheinberger 1992), how embeddedness in social contexts shapes the ends and meaning of technological objects (*causa finalis*) (i.e. Kopytoff 1986; Pinch and Bijker 1987; Daston 2000; Rogers 2003), or how individual creativity manipulates the material and social world of artifacts and leads to inventions (*causa efficiens*) (i.e. Joas 1992; Schumpeter 2000). However, such approaches fall short of explaining how artifacts open up to meaning-making because they work, because they function in a manner that exploits the capacities of their elements due to their allocation (*causa formalis*).[25] This is, for example, how a solid and flexible chain adapts to a round sprocket, while still being resistant enough to transduce the energy that drives a hub. Taking form seriously argues that allocation of elements constitutes an object's mode of existence. This is, for example, arguing that concretizing the allocation of chain, sprocket, and hub into a functional unit opens this unit up for applications such as the drivetrain of diverse vehicles.

Hence, stressing form focuses on a technology's material becoming. Ernst Cassirer regarded stressing form as shifting focus away from technology as a product toward its mode of producing and the structuring forces that are revealed through it (Cassirer 1985, 49). He urged the understanding of technology from its coming-into-being; the form not only expands agency, but changes agency in a qualitative sense, which entails the potential to reveal a new aspect of reality (Cassirer 1985, 53). His understanding of form is active and might be comprehended through its verb *forming*, or process *formation* (cf. Hoel and Tuin 2012, 192). In that sense, he regarded technology as similar to language, as both were means to accommodate and construct the world: language through lingual-theoretical thought and technology through its material functioning. Hence, giving materialities a technical form is equally assembling and revealing the world. This thinking is similar to Simondon's *concretization*, which he regarded as individuating towards a technical essence and becoming agential or effective in the world (Simondon [1958] 2012a, 19ff.). The technical form, in that sense, is a shape less defined by a specific morphology and more by the relations that are stabilized through it. Based on Cassirer and American pragmatism, Rammert understands the technical form not as being defined by its matter, but by its efficacy in stabilizing relations. The technical form is a situatively found, tested, perpetuated, and repeatable sequence of activities with predictable outcomes that can materialize through symbols, matter, and bodies (Rammert 1998, 308). Based on these remarks, I regard a technical form as a functioning and stabilized set of relations whose elements become effective through their sequential order. In relation to novelty, this brings about the two relevant questions: *what does a technical form articulate, and what is novelty in this regard?*

Still, I want to emphasize that asking how a technical form evolves is an analytical focus that I regard as neglected; it is not meant as explaining form through formation or seeking other determinisms. In this regard, I see agency as a pivotal

---

25   Some might argue that form and formation are major concerns of Actor-Network Theory (ANT). That is right, but not in terms of novelty. In ANT-studies, processes of form are usually considered as delegating action to material entities; this is rather related to questions concerning power and less to expressing novelty (cf. i.e. Latour 2005; Law and Singleton 2005).

concern. Generally, inquiring into technical form empirically suggests symmetrical approaches of human and non-human agency, as proposed by scholars influenced by Actor-Network Theory (ANT) (i.e. Latour 1987; Akrich 1992; Mol and Law 1994). Their empirical perspective regards material entities as agents in constructing the social world; they bring in the "missing masses," as Latour would say (Latour 1992). However, other scholars who have urged material agency, too, have stressed that the semiotic symmetry of ANT dead-ends when it comes to explaining "intentionality" (Pickering 1995, 17), like plans and anticipating scenarios, or processes of "sense-making" (Rammert 2008, 8) that give meaning to technologies and artifacts. Simondon, whose philosophy somewhat pre-empts ANT's trademark acknowledgement of material agency, also stressed that concretizing technological objects requires an inventive anticipation that organizes the relation of technical entities that only exist after constituting an object. In that sense, anticipative thinking conditions the present through the future, as it imagines the structural conditions through which new technical forms evolve (Simondon [1958] 2012a, 53). Regarding the four causes, this means that materiality, meaning, and human action take part in the genesis of a technical form, and each with different impacts. However, in the style of ANT-scholars, I regard their agency as an empirical question addressed from their enactment and impacts and not from their *a priori* status as a category of existence, like in Aristotelian teleology or anthropocentric social theories. This includes how agency is distributed in hybrid constellations (Rammert 2008; Rammert 2012) and how their heterogeneity stimulates the processes that give an object its form.

As outlined, my understanding of form is active and concerned with processes of becoming. This places the focus on the openness of processes through which new forms articulate novelty. However, nothing is totally open or naïvely new. There are conditions, such as technical requirements and histories of technologies that precede objects. In the previous section, I address such kin objects as the generalized collective to which an object identity relates, like the collective of robotic hands for the RBO Hand. The focus of that section is discursive processes that figure an object's meaning. In this section, I want to introduce a similar notion that captures how an object becomes one of a specific type, but, instead of discursive relations, I regard specific technical features as typical for a type of object. These features are part of the form, but, instead of emerging as an object's novelty, they are elements that continue through different objects of a type and constitute a shared understanding of what design features commonly belong to objects of that type. I consider the sum of such features a *format*. The format defines what features a form needs to embody in order to become a specific kind of object. Hence, the format is a trajectory of stabilized technical relations that shapes a group of objects. In the following, I describe two of such formats: robotic hands, which is the format of the RBO Hand's form, and cybernetic machines, which I regard as the format of *Mirage*. These formats define the technical features that I bring along to compare the forms of the RBO Hand and *Mirage*.

An empirical challenge of this paper is the selection of which object state is stable enough for a comparison of forms. Whereas the genesis of *Mirage* proceeded toward a "natural" empirical climax, which is its first exhibition requiring stabilized functioning, the RBO Hand does not have such an endpoint for its development. Even public enactments through research papers stress the Hand's preliminary status. This difference indicates the sampling problem of this ethnographic comparison: although both technologies share similarities, their designs follow very different cycles. Hence, it makes little sense to compare both objects' forms in a step-by-step description of the design work, which would also demonstrate bias toward a *causa efficiens*. Instead, I consider the accumulation and stabilization of technical features as sampling criteria and as a comparative structure. In return, deciding what stabilization is sufficient to be regarded as form can be linked to decisions made by the actors: for the RBO Hand, that reference state is when the Hand functions with enough stability to do experiments with publishable results, and for *Mirage*, as has been said, that state is its first exhibition. These states are the reference states for analyzing the novelty of both objects' forms.

I begin the analysis by sketching the two formats of robotics hands and cybernetic machines. From there, I turn to the RBO Laboratory and Baecker's studio and describe the hybrid constellations of material inquiries. That section addresses the problem of agency mentioned above. In the following section, I will compare the RBO Hand and *Mirage*'s form. The comparison is structured according to the technical features, which I have summed up as formats. Concluding this chapter, I address the relation of novelty, form, and hybrid constellations before I close with a detour into two techno-aesthetics.

3.1 Formats and Their Technical Features

In the following, I sketch typical technical features of the formats of robotic hands and cybernetic machines. The formats capture the technical structure, which is typical for such objects and their engineering. Comparing the relation of the RBO Hand to the format of robotic hands with *Mirage*'s design and its relation to cybernetic machines is somewhat problematic. The problem is in assigning the RBO Hand and *Mirage* a specific format. Whereas assigning the RBO Hand to the format of robotic hands is easy because the typical features are easy to identify and the actors label the Hand as such, assigning *Mirage* a format is an interpretative task. Interpreting *Mirage* as a cybernetic machine is ambivalent, as

Baecker referred to an artificial neural network as a related technology and did not mention cybernetics explicitly in relation to *Mirage*. In that sense, categorizing *Mirage* as a cybernetic machine is, by my account, based on similarities in their technical features. However, I have to admit that this account is not only my own analysis of how far *Mirage*'s technical features were previously embodied in cybernetic machines from the 1950s and 1960s. Baecker pointed me in the direction of interpreting *Mirage* as a cybernetic machine, by mentioning Ross Ashby's *Homeostat* in a presentation that he gave about his general artistic approach. Nevertheless, that presentation came prior to the development of *Mirage* – and, more crucially here, was a discursive reference similar to his enactment of artificial neural networks akin to *Mirage*. Where the focus is here, in contrast, on the technical features that several objects share and that the RBO Hand and *Mirage* have embodied and/or contested. In that sense, the following features map the typical technical structures of both formats.

3.1.1 Technical Features of Robotic Hands

Paying attention to a robotic hand's technical structure makes the fragmentation of robotic grasping as a research field comprehensible. In robotic-hand literature, this field is segmented into functional bits that enable research to be precise about what problem is being contributed to. Controzzi et al. wrote a review on the design of artificial hands, for which they did not consider their development as a linear evolution as did Rosheim, but in terms of how different design features of robotic hands developed. They pointed out five[26] issues that the design and development of robotic hands should consider (Controzzi, Cipriani, and Carrozza 2014, 225ff.). These rather abstract issues, which also appear in several other technologies, can be used for structuring the typical technical features of robotic hands.

1. *Kinematic architecture*: The kinematics of a hand capture the choice of (controlled) degrees of freedom as well as the numbers of joints and actuators to trigger the movements. Thus, it directly influences the performance of a hand. For instance, the Salisbury Hand consists of nine degrees of freedom, nine joints (three on each finger), and twelve actuators, a number perceived as the minimum for achieving dexterous manipulation. The higher number of actuators compared to its degrees of freedom increases the potential dexterity of the hand. However, the choice of more actuators than joints is a trade-off, as it also yields to bulkier and more complex systems and controls. To tackle this issue, systems with cou-

---

26 In the original article, they point out six, not five, features, but I will address "anthropomorphism" in other sections of this study.

pled transmission and "underactuation" have been developed (Controzzi, Cipriani, and Carrozza 2014, 226-7). A kinematic architecture is generally referred to as underactuated when the system has fewer inputs (controlled signals) than outputs (degrees of freedom). This design principle employs differential mechanisms, mechanical limits, and elastic elements. The advantage is automatic adaptation to the shape and contour of items that are to be grasped, and therefore an increase in contact area and, hence, stability. An early but still representative example of underactuated mechanisms is the soft gripper developed by Shigeo Hirose, which is able to softly and gently conform to objects of any shape and hold them with uniform pressure.[27] One modern example is the SDM Hand, whose authors emphasize that it performs a wide range of grasps by using only a single actuator (*Figure 10*) (Dollar and Howe 2010).

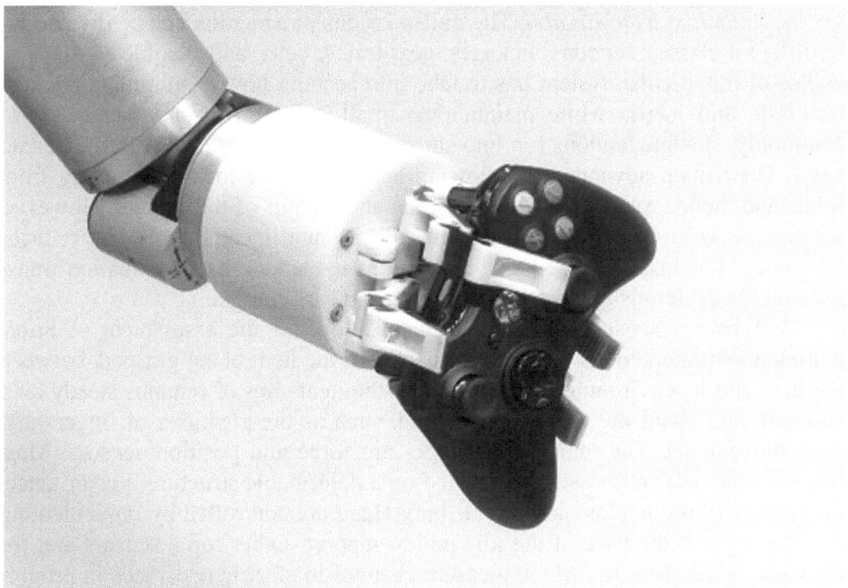

Figure 10: The SDM Hand (source, Dollar and Howe 2010).

2. *Actuation principle*: The actuation principle specifies what is used to do the work that muscles do in human hands. The main actuation principles are DC

---

[27] The system was inspired by snake movements. Its inner workings can be comprehended here: http://www-robot.mes.titech.ac.jp/hirose/robot/snake/sg/sg_e.html (last accessed September 5, 2014).

motors, pneumatic or hydraulic valves, and shape memory alloys. Most robotic hands are actuated by electrical motors. Their main advantage is their precision and ability to store electrical power in small batteries. Pneumatic actuators, however, have the advantage of showing inherent compliance caused by the fluidity of the material used for power transmission. This compliance is an advantage in terms of safety during human-robot interaction, "but, since it is difficult to modulate, it becomes a disadvantage during the execution of precision tasks" (Controzzi, Cipriani, and Carrozza 2014, 228). Furthermore, pneumatic actuators allow integration of the actuation transmission with the fingers, which makes possible compact and lightweight designs (Schulz, Pylatiuk, and Bretthauer 2001). An important index for comparing actuation principles is the power-to-weight ratio of a system, which provides an idea of its power density and is useful, especially when lightweight solutions are being sought.

3. *Actuation transmission*: The transmissions of actuation can be divided into different classes: tendons, linkages, gear trains, belts, and flexible shafts. The choice of a particular system has to take into account how to minimize friction, backlash, and inertia while maintaining small overall size and weight. Most commonly, flexible tendons run into sheaths, analogous to tendons in the human hand. Their main advantage is allowing actuators to be located remotely from joints and, hence, reducing the dimensions and weight of the fingers. However, friction between tendon and sheath introduces non-linear effects and reduces efficiency. For instance, the Salisbury Hand uses tendon-based actuation transmission for dexterous grasping, but suffers from early fatigue.

4. *Sensors*: Sensors in a robotic hand allow for the assessment of information about interactions between the hand and the item being grasped, between the item and its environment, (e.g. whether the item slips or remains steady on a surface), and about the status of the hand, such as the positions of fingers and joint movements. The main sensor types are force and position sensors. Most force sensors use strain gauges mounted on a deformable structure. For instance, the tension of the tendons in the Salisbury Hand are controlled by implementing strain gauges at the base of the idle pulley support. Other force sensors are, for example, tactile sensors, which measure changes in electric resistance in proportion to pressure on the top and bottom of a thin film that is implemented on fingertips and palms. Position sensors are usually indirect measurements, such as optical encoders, which inform the system about current finger configurations and positions (Controzzi, Cipriani, and Carrozza 2014, 234-5).

5. *Materials and manufacturing method*: Like any other mechanical objects, the choice of material affects most of the features of a robotic hand, such as its weight, compliance, or strength. As in all engineering fields, there are many factors taken into account for the choice of materials and their manufacturing

processes. Constraints for the choice of materials in robotic hands are, for instance, the range of possible wall thicknesses or corrosion. Robotic hands are usually assemblages of different materials that vary among components. In state-of-the-art hands, compliant materials are becoming popular. One of their advantages in the manufacturing of joints is that they avoid a number of components, such as pulleys, axes, or torsion springs, which results in a reduction in joint sizes. Moreover, on the surface of a robotic hand, compliant material possesses specific capacities that are investigated analogous to human skin. The force-deformation characteristic of human skin on fingertips plays a fundamental role during precision grasping and manipulation. The usage of new materials is also encouraged by the introduction of new manufacturing processes that allow rapid prototyping. Whereas design and fabrication of a prototype using traditional machinery techniques is a rather long and expensive process, technologies like 3D-printers allow materialization of a conceptual design rapidly and at low costs.[28]

3.1.2 Technical Features of Cybernetic Machines

As mentioned, *Mirage* is not as clearly a cybernetic machine as the RBO Hand is a robotic hand despite its deviance in several features. The following features are significant for cybernetic machines and map their technical structures. However, mapping these features is problematic, as cybernetic machines are primarily labeled as such due to their philosophical or epistemic value and not their shared technical features. Furthermore, machines developed in cybernetics are technically more diverse than robotic hands, which have a stable functional core (grasping) across the range of different hand designs. Thus, the following features are aggregated from different technical (Müller 2014) and historical (Hayles 1999; Pickering 2002) sources, and I consider them typical features of cybernetic machines as embodied in some of such machines most prominent examples. Nevertheless, the listed features are highly selective and guided by my prior knowledge about *Mirage*'s technical structure. In that sense, these features are what you (the reader) need to know about cybernetic machines in order to comprehend how far specific technical features have a history as a format.

1. *Embodiment*: Stressing the materialization of machines is not a tautological discussion for cybernetics. That is, cybernetic machines are commonly embodiments of abstract concepts that claim to be generalizable similar to theories that could also be made intelligible through mathematic symbols or language. Howev-

---

28 Compare, for instance, the Open Hand Project that aims to make robotic prosthetic hands more accessible to amputees: http://www.openhandproject.org/ (last accessed September 10, 2014).

er, strands of cybernetics that are particularly concerned with technological machines seem to have a more significant relation to *building* machines that embody their concepts.[29] A historical example for this relation is a letter by Alan Turing, in which he proposed to cybernetician Ross Ashby that Ashby should run his "special machine," which later became known as the "*Homeostat*" (*Figure 11*), on Turing's *Automatic Computing Engine* as a simulation.[30] As history has shown, Ashby declined Turing's offer and built his special machine as an embodiment of what he called an "ultrastable system." The turn to building machines instead of an epistemology of symbols has encouraged Pickering to acknowledge cyberneticians like Ashby as ancestors of his Mangle of Practice. According to Pickering, the machines built by Ashby, Stafford Beer, and Gordon Pask are "all about this shift from epistemology to ontology, from representation to performativity, agency and emergence" (Pickering 2002, 414). This turn to agency through embodiment is similar to *Mirage*. As mentioned above, *Mirage* is narratively related to artificial neural networks, which are, first of all, algorithms based on symbols. For *Mirage*, in contrast, Baecker attempts to give body to the abstract concepts of such learning algorithms. In that sense, he refuses to continue with his "hallucinating machine" as algorithmic symbols, but turns to a material embodiment of such concepts, similar to Ashby's reaction to Turing's offer.

2. *Input signals*: The pivotal topic of cybernetics between the 1950s and 1960s was homeostasis (Hayles 1999). Homeostasis is the ability of organisms to maintain a steady state in disruptive environments. Cybernetic machines perceive environments as input signals. Signals can be: electronic signals like electricity, voltage, and induction; mechanical signals like pressure, torque, or acceleration; other signals like temperature or light. Considering input signals as a technical feature of a cybernetic machine stresses that its design determines what signals are actually perceived by the sensors and, hence, what the environment actually is. Ashby's *Homeostat* is based on electric input signals that pass through a coil that generates a magnetic field, which makes a needle rotate in a specific direction. Nevertheless, input signals that stimulate cybernetic machines are not any kind of electric, mechanical, or other signals. What makes them significant is that input signals are not stable or determinate in their frequency and amplitude. Rather, cybernetic machines are stimulated by input signals that are random and unpredictable. In Ashby's *Homeostat*, this is shown in the varying current stimulating the needle. A better example, however, is the varying degrees in temperature, which make a thermostat a sensible device.

---

29   Of course, turning to embodiment is significant only for certain strands of cybernetics. For others, the opposite is even true. Hayles, for instance, has stressed that the cybernetic figure of the human being is dramatically bodiless (Hayles 1999, 4).
30   Letter by Alan Turing to Ross Ashby from November 1946 and source of *Figure 11* accessible via the Ross Ashby Digital Archive: http://www.rossashby.info/ (last accessed: March 4, 2015).

3.1 Formats and Their Technical Features

Figure 11: Ross Ashby's Homeostat (source, The W. Ross Ashby Digital Archive).[30]

3. *Converter*: After sensors have perceived random input signals, they are converted within the main technical body of cybernetic machines. What I have simplified to call "converters" here are actually what require the most inventive effort of cybernetic machines. I use the term in a very broad sense, meaning a technical device or system for altering the nature of an input signal and passing it on toward an output. Hence, converters are functional units for manipulating and transmitting signals, within which input and output and their temporal effects are placed in relation to another (Müller 2014, 6). A thermostat is a converter that aligns varying temperature with a predefined target value; their difference manipulates the heating system to regulate its state. Similarly to input signals, many technologies implement converters, for instance, to convert analogue into digital signals, or AC into DC. In that sense, converters embody a main concept of cy-

bernetic machines, that is, *feedback loops*. In cybernetics, feedback loops are the flow of information between an organism and its environment.[31] Long before the advent of cybernetics, the centrifugal governor, engineered by James Watt in 1775 and from then on used in steam engines to regulate steam flow and stabilize their output, is already a regulation device based on feedback loops. In cybernetics, the same concept became matter of theory-building and the focus of the discipline's constitution (Hayles 1999, 8). Feedback loops regulate information flows not only within systems, but also between system and environment.

4. *Output signals*: Considering output signals now is a logical step, but also requires a few words to explain what makes them significant for cybernetic machines. Firstly, converters regulate and manipulate output signals. Equally, output signals of a mundane thermostat or those of Ashby's *Homeostat* are signals that are aligned with the condition of their environment. Secondly, output signals can be either of the same signal type as input signals or transformed to another type of signal. The thermostat and *Homeostat* are both examples wherein the signal type does not change; signals remain either as temperature or electricity. In contrast, the aesthetic devices of Gordon Pask transformed signals into another signal type. For instance, his *Musicolour Machine* is based on a feedback loop running from the human performer through a musical instrument that converted electricity into sound and light, which was then fed back to the performer (Pickering 2002, 427).

3.1.3 Formats and Their Core Functionalities

The relation of format and form is different in both objects' becoming. The RBO Hand is closely related to the common features of robotic hands, as these capture essential parts of the shared engineering knowledge in the field of robotic grasping and are issues that can be contested through alternative designs. The features list the common technical necessities for engineering the functional core of robotic hands, which is *grasping*. Grasping is the basic functionality that structures any robotic hand's material form. The RBO Hand can explicitly contest such features in order to make its novelty comprehensible. For example, research could explicitly aim at rendering sensors obsolete or propose a more efficient actuation transmission.

This is different for *Mirage*. The design features of cybernetic machines are not common knowledge in the field media art, to which *Mirage* most likely contributes. Hence, it is unlikely that technical details legitimize *Mirage*'s novelty.

---

31 For Nobert Wiener, feedback loops marked the end of the liberal subject (Hayles 1999: 2).

However, mapping the format of cybernetic machines entails an assumption: there is something comparable to the functional core of grasping, and this functional core is a point of convergence in the assemblage of *Mirage*'s technical components. Listing the technical features of cybernetic machines indicates that this core is their significant object-environment relation. This relation entails technical features that sense and convert random signals into an output. Similarly to grasping, I consider *converting* as a technical functionality that signifies the format of cybernetic machines.[32] To what extent the converting of signals is a basic functionality that structures *Mirage*'s genesis similarly to grasping in terms of the RBO Hand is a question of this section.

From both lists, I sum up that formats define basic functionalities realized through specific technical design features. The formats described define what a technical form needs to embody in order work as a specific type of object. They define technical relations inside an object's form; they are schemes of functionalities. Actors can appropriate these as knowledge about how technologies work in general. They allow for the signifying of a form as akin to a specific object type or, on the other hand, for the deviation in specific aspects from that format in order to signify their difference.

3.2 Hybrid Constellations, Inquiries, and Distributed Agency

Formats are not just knowledge, nor are they absent from spaces where a form evolves. They are inscribed in technologies, tools, consumables, software, etc. in the laboratory and studio, the places where a new form comes into existence. This is, for instance, how infrastructural technologies "pre-scribe" (Latour 1992)[33] the range of shapes a robotic hand may enter, or what technical means are at hand. In the following, I address the constellation through which the RBO Hand and *Mirage* enter a specific form, as well as the inquiries that enact its agency. I address these as hybrid constellations of people, machines, and programs. I follow Rammert in this regard, and consider a constellation as "the mode how agencies of heterogeneous instances are distributed and connected with one another and the level of agency that is given to them in certain situations" (Rammert 2008, 16). This understanding considers material, human, and

---

32 Of course, I do not consider converting as a functionality that signifies cybernetic machines as exclusively as grasping signifies robotic hands. However, it is a technical similarity between many machines engineered in the cybernetic paradigm (cf. Pickering 2010). Surely, one could also mention, for instance, "adaptation," as Pickering did, but I have chosen "converting," as it has a more technical and active connotation.

33 Latour understands prescription as the "behaviour imposed back onto the human by non-human" in its moral and ethical dimensions (Latour 1992: 232).

semiotic entities as potential agents, but, in opposition to the flat symmetry of ANT, allows different levels of agencies, depending on an entity's situated enactment. It is similar to what Suchman has regarded as "configuration" (Suchman 2012), as it delineates the boundaries of an object. However, in contrast to her notion, constellations focus on the causal efficacy of agencies instead of how they narrate the significance of their existence.

Some entities of such constellations are already mentioned in Section 2.1.2, where I briefly report on the laboratory workshop and the studio and what material traces of past inquiries I observed there. There, I report on some of the necessary tools and consumables for manufacturing both objects. These influence both objects' forms, as they enable and limit specific tinkering practices. In the following, I focus on entities that are technically more advanced. Some of these enable specific tinkering practices, too, whereas others impose both objects' potential existence. Similarly to social conditions, descriptions of hybrid constellations can hardly capture all possible entities that effect a form; there might always be hidden agents influencing an object's design.

3.2.1 Robotics Infrastructure and "Everyday Life" in the Laboratory

One way to distinguish between different approaches in robotics research is by stressing whether a robot is designed as a whole or modularly – that is, when all its parts are developed as a holistic project, or if an approach aims at developing separated parts like legs, sensors, or hands. Whereas the first requires large-scale projects with long-term funding and, therefore, needs to be concerned less with fitting with existing standard robot technologies, the latter focuses on problems of a given domain and, hence, relies on standard hardware to fulfill all those robotic tasks that are not issues of the investigative effort. It is rather obvious that the RBO Hand belongs to a modular approach, as everything that I have reported on so far refers to the domain of robotic grasping. Nevertheless, labeling the RBO Hand as a modular approach is only appropriate depending on the point of view. As mentioned above, robotic grasping is a very fragmented domain and, hence, developing a whole hand could be regarded as a holistic approach. In that sense, the RBO Hand's approach is holistic, because it covers kinematic architecture, actuation, sensing, and material – which could be addressed on its own to make a research project. However, the Hand is also modular, as it is designed to fit a standard robot arm.

This standard robot arm is part of the laboratory's scientific infrastructure. It is a Meka Robotics A2 robotic arm with seven degrees of freedom. The arm is attached to a robotic torso, the Meka T2. Meka is a US company that manufactures robot hardware and software especially for robotics research. When applied to grasping experiments, the arm defines the movements with which a robotic hand approaches an item. In research papers, the arm is explicitly mentioned as part of the experimental setup (Deimel and Brock 2013, 2043). According to the company's brochure, the arm matches the size and shape of a small adult, which makes it an ideal platform for researchers interested in the manipulation of human environments. It features force-controlled actuators, intrinsic physical compliance, zero-backlash gearheads, and standard software for manipulation control. These features are important to mention, as they influence the arm's behavior in experiments. According to the brochure, the arm provides plug-and-play support for the Meka T2 Humanoid Torso, the S2 or S3 head, and the G2 or H2 hand. Its actuation principle is declared as a "Series Elastic with Torque Control." Whereas the upper end of the arm is fixed to the Meka Torso, the lower end allows for the attachment of different robotic hands. Obviously, the company also provides its own hand, which is, for instance, used by the researchers, who are only interested in planning tasks. Nevertheless, Meka also provides a mount for attaching different manipulators. Like a human arm, the Meka A2 has as flexible shoulder, elbow, and wrist. These three joints work together to produce an arm movement. Important to note here, the flexible wrist belongs to the arm.

The Meka A2 matches the format of robotic hands. It considers various technical standards and potential hand shapes. In that sense, it also influences and pre-defines aspects of the RBO Hand's form in four regards: Firstly, the Hand's general size should be somewhat aligned with the measurements of arm and torso. This does not need to be a precise match, but the ratio should allow the assembly of Hand, arm, and torso to move in all possible directions without being disturbed by the Hand's body. Secondly, the Hand needs to be mountable onto the arm. This requirement is not as banal as it sounds. For instance, mounting the Hand onto the arm enables Deimel to run controlled experiments with the hand, something the Starfish Grabber and the other preliminaries cannot perform. Thirdly, the Hand needs to have a precise weight in order to define what force a robot arm has to deliver to fulfill controlled movements. This weight has to be represented in the steering software of the robot arm. If the weight is incorrect, either the arm will lunge out or move impeded. Fourthly, the range of arm movements allows the hand to limit its movements to closing its fingers. That is, the Hand's design can focus on grasping through finger bending. In that sense, the design can rely on the arm to fulfill all other movements necessary for grasping an object, such as moving toward an object and positioning the Hand by moving the wrist – given that the arm works as expected.

However, the arm does not always work as expected. The video sequence in *Figure 12* shows the inquiry of creating the setup of the Meka robotic arm and the RBO Hand work together. The sequence is a mundane situation in the RBO Laboratory. It shows Deimel preparing the setup for a test run. He has attached the RBO Hand onto the wrist of the Meka arm and positioned a table with a bottle in front of it. The arm does not work as expected; its movements differ from those required for an experiment. According to Deimel, the weight of the Hand causes the malfunctioning, as its entry in the steering software is incorrect. This brings about deviant arm movements. However, entering the corrected weight at one position does not simply solve the malfunctioning. The arm has several degrees of freedom whose forces require the precise alignment of parameters, because weight/force ratios differ depending on the kind of movement. This leads to trial and error work. As the sequence shows, this work is coordinated through several agencies. Firstly, Deimel's activities focus on the steering software. In the software, the Hand is represented in coding symbols. The deviating arm movements bind the symbols to the material world. Here, code is not an alphabetical or arithmetic sign system detached from materialities, but rather, it works as the grammar that conducts the ensemble of the robot arm, Hand, and software. Deimel pulls the table away so the arm does not run into it. With the table pulled aside, Deimel can see how the arm moves without being disturbed through its environment. After the arm has driven into its default position (Frame II), Deimel tries to access the system state. In Frame III, he uses his body to do so and pushes the robot arm. By pushing, he triggers a movement, which indicates an active system state. He cannot tell from the information on the monitor alone if there is tension in the arm's joints; he needs to feel it. In Frame IV, Deimel first recodes the software. He anticipates that the arm might still behave unpredictably and puts his left hand on the emergency button, which allows him to shut off the movements immediately. After pressing enter, the arm makes a complex movement using several of its joints. In total, the sequence illustrates that it takes human bodies, physical things, and algorithmic signs to constitute technology (cf. Rammert 1998, 317). This is not only true for operating machines, but also for revealing a new form that needs to fit into a technical structure.

Infrastructural technologies like the Meka A2 are not commonly found outside laboratory or research settings, as they are developed as scientific tools. In that sense, the Meka robotic arm is a very specific technology that stems from the Hand's origin in a scientific context. However, not only sophisticated technologies are part of the laboratory's infrastructure. When visiting the laboratory, there were clear indicators that grasping experiments take place there. That is, there were several mundane items lying around – not only specifically placed like the bottle on the table shown in the sequence, but also randomly distributed on desks. One of such items appears in Section 2.2.3: the apple that was used by

3.2 Hybrid Constellations, Inquiries, and Distributed Agency 123

	Time	Still of the video	Transcript
I.	04:28		Deimel programs the steering software of the Meka A2. He looks at the table and pulls it out of the Hand's reach.
II.	04:37		He continues programming. After writing, he hits the keyboard like pressing Enter. The Meka A2 drives downwards and stops parallel to the robot torso.
III.	04:55	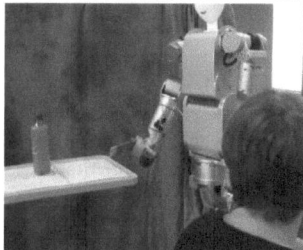	Deimel looks at the new position and pushes the Meka A2 slightly with his left hand. The robotic arm moves outwards and swings back to its original position. Its movements are constrained through its active system mode. It behaves similar to a human arm with contracted muscles.
IV.	05:08-05:27		Deimel orientates back to the monitor. He writes code, presses enter again and puts his left hand on the emergency button. The Meka A2 moves backwards. Then it turns slightly around its axis and moves upwards. It stops in a 90° angle to the torso. He continues programming.

Figure 12: Inquiry patterns in the RBO Laboratory (own video).

Deimel to enact the Hand's compliant material. That apple was not a natural but an artificial apple. Hence, its only purpose in that space was to serve as an item to be grasped by the Hand. Beside the apple, there were several other items lying around, like a bottle of juice, which had its label peeled of, and a tennis ball and a tub. These items appeared somewhat out of place, as they did not seem to belong to a laboratory setting. However, they played a crucial role in the conditions for the Hand's design: they embodied "everyday life." They brought into the laboratory the figured context of a robot's future use. In that sense, they embodied the imaginary of an autonomous robot, as their presence became a meaningful signifier of a humanlike machine (cf. Suchman 2007). Nevertheless, this is not the argument that I would like to push forward here. Certainly, the presence of everyday items figures the humanoid robot, but despite embodying imaginaries, these items had an immediate material impact on the RBO Hand's design: they were the contact surfaces used to optimize the Hand's grasping abilities – as easy as that. However, this is not trivial. For instance, there were also sheets of paper lying around, not normal paper but blotting paper (for whatever reason). Sheets of paper are everyday items that are extremely difficult for common robotic hands to grasp, as they are (almost) not three-dimensional. The robotic hand would need to be able to use the table for resistance in order to pick them up. Unlike other hands, the RBO Hand is rather good at this, because its inner layer is sticky. This allowed the Hand to push the paper against the table, cramp it, and pick it up – a technique that was tested and optimized in the laboratory setting. In that way, the everyday objects that lie around in the laboratory are significant to the constellation, as they are articulated in the Hand's form; they make the Hand's distinctive grasping style become effective and observable, and, following, influence what technical pragmatics are advanced as novelty.

3.2.2 Anticipated Conditions and Inquiries with Open Hardware in the Studio

The constellation of *Mirage*'s becoming is considerably different. In contrast to the RBO Hand, *Mirage*'s design does not need to consider any standard infrastructural technologies like the Meka robotic arm does for its future existence. In comparison, *Mirage* is a standalone technological object that relies only on electricity to technically function when finished. However, several future conditions require consideration in *Mirage*'s design. Firstly, Baecker has to consider the spatial conditions of potential exhibition spaces. As *Mirage* is not Baecker's first media installation, he can tell from several past experiences that one needs to be anticipating an exhibition situation in the early stages of a project. In Section 2.1.2, I report on a structure for testing how mechanical pulls translate between materials, such as between wire and a mesh of strings. Within the same situation, Baecker told me that one reason for such testing is that he needs to consider transmission ratios when he wants to work with mechanical pulls. He said that it might well be that a pull translates through different pulleys and connections with a ratio that increases the movement quite dramatically. Hence, he cannot simply assume that any mechanical idea is realizable for an installation, as exhibition rooms offer only limited space.[34] Secondly, the exhibition room needs to provide appropriate lighting conditions. In that respect, Baecker can also tell from experience that optimal perception of his installations requires dark environments. Especially when working with light or projections, it is only feasible if one can assume that an exhibition room can be darkened properly or painted black. The availability of dark exhibition rooms is reasonable to assume, said Baecker, although one will never have perfect conditions. Certainly, he added, one always needs to expect changing conditions in a location. Many of my ethnographic visits in his studio took place in the evening when it was dark, or Baecker blocked the windows with blankets to simulate exhibition conditions.

Similarly to the laboratory, there are several technologies in the studio, too. Differently from the RBO Hand, they are not necessarily technical infrastructures. Several technologies in the studio are tools, ranging from drills to computers. I list some of these in Section 2.1.2, where I describe that materials and test settings become traces of previous inquiries. Several technical tools are necessary for building *Mirage* and most of Baecker's previous installations, but not all of them are significant, as they are for general purposes. Others, however, are more significant, as they are for specific purposes and have a direct impact on the technical means and media of an installation. What I would like to mention as a significant technology, analogous to the Meka robotic arm, is Baecker's use of open hardware. The availability of open hardware such as Arduino circuit boards[35] has a tremendous impact on Baecker's work, as it allows him to integrate computational components into his installations and couple these with mechanical elements. For instance, Arduino circuit boards have several digital and analogue interfaces and a processor that allows easy programming and flexible connections. A basic Arduino board has 14 digital input and output pins, six analogue inputs, a 16MHz ceramic resonator, a USB connection, a power jack, an ICSP header that allows in-system programming, and a reset button. On the website of the open hardware initiative, there are plenty of more addable, digital, and analogue hardware components for sale. The open hardware shares with the Meka robotic arm that it enables specific forms. In contrast to the Meka robotic

---

34 Baecker's story can be well comprehended by looking at his installation *Rechnender Raum*, for which he combined ancient mechanics with digital technologies. The installation has a diameter and height of approximately two meters. It translates several mechanical pulleys into a complex movement of strings.
35 Website for Arduino: http://www.arduino.cc/ (last accessed: March 11, 2015).

arm, open hardware does not allow for limiting technical relations, but enhances the range of potential technical relations within a form.

The diversity of materials significantly structures Baecker's artistic practice. The use of open hardware and the opportunity to substitute and interlink digital and mechanical technologies structures his inquiries for realizing the "hallucinating machine" as well. Such tinkering is shown in the comparison of how he enacts the materialities of the test structure mentioned in Section 2.2.3 with a later test structure for which he used an Arduino board and elements of acrylic glass. The transcript in *Figure 13* starts shortly after *Figure 9*. The aforementioned test structure consists of two wooden plates, hooks, threaded bars, strings, nuts, and bolts, as well as customized pulleys and an elastic element (*Figure 14*). Baecker is setting up the string, as he wants to test how it moves when threaded through the hooks and pulleys. He does not thread the string through all of the hooks at once, but stops after a few rows. He stands upright and pulls the strings several times with changing intensities. The elastic element, which is located at the fixed end of the string, and the number of hooks through which the string is run, determine the string's behavior. Baecker does not simply pull the string in order

Time	Still of the video	Transcript
I. 05:10		Baecker threads the line through three hooks and pulls it several times softly. The line pulls back every time due to the elastic element. He threads the line through another hook, stands upright and pulls the line ten times. Every pull is slightly different from the one before.
II. 06:15 – 07:00		Baecker keeps on pulling the line slightly. Baecker: *"There is already too much friction on it; it won't work that way."* He stops pulling and starts to decoil the line. Baecker: *"It is already too tight."*

Figure 13: Baecker using his body as actuator and epistemic tool (own video).

3.2 Hybrid Constellations, Inquiries, and Distributed Agency 127

Figure 14: Test structure made from wooden plates, hooks, and strings (own picture).

Figure 15: Test structure made from Arduino board, bows of acrylic glass, and wires (own picture).

to trigger the movement, but uses his body to feel how the string behaves depending on the force he applies to it. In Frame II, he concludes, there is already too much friction on the string, which makes the setup insufficient for his use. Either he has to change single components or think of a different structure. The sequence shows how Baecker is using his body to feel the string's behavior; he enacts the components and accesses their relation. Similarly to Deimel, who uses his body to access the system state of the robot in *Figure 12*, Baecker's body becomes actuator and epistemic tool. When I visit him a couple of weeks later, the test setting has changed drastically (*Figure 15*). The previous structure is placed in one of his upper storage shelves, and Baecker works on a small structure of acrylic glass. The new structure consists of a platform with five attached bows. Between the platform and upper end of each bow is a wire. All the wires are connected to the analogue outputs of an Arduino board. Baecker tells me that it is a board he had lying around, but he actually wants one with additional input jacks. Through the board, he is able to actuate the wires and trigger a patterned movement. Although the materials have changed completely compared to the previous structure, *movement* remains the focal point of Baecker's inquiries. In that sense, the associative reference of a "wafting area" also continues in the new test structure. Instead of physically actuating a string, Baecker now substitutes his body for an Arduino board that actuates the structure through electronic signals.

3.2.3 Distributed Agency

The constellations described above in the RBO Laboratory and Baecker's studio stress the problem of agency, which I address by introducing this chapter. With agency, I refer to the basic capacity of entities to cause effects. The distribution of agency stresses, for instance, how humans delegate actions to technologies, and how these refuse such delegations or, in return, impose behavior back onto humans. A new form evolves within such messy hybrid constellations, interfacing heterogeneous instances of different agency levels. Rammert distinguishes between three levels of agency: causality, contingency, and intentionality. *Causality* refers to the basic ability of agents to cause effects in the world, *contingency* means the capacity to act in a different way, and *intentionality* refers to the domain of meaningful action (Rammert 2008, 10-1).[36] Whereas these criteria commonly constitute human action in social theory, the notion of a constellation stresses the question of how material and sign-based entities receive comparable agency levels depending on their positions and enactment. I want to summarize some modes of enactment from the preceding empirical descriptions that capture the hybridity of relations.

The first mode of enactment regards the *position* of the RBO Hand in a robotic ensemble. This is the requirement of fitting the Hand into the Meka robotic ensemble and supporting and building upon its technical agency. The Meka arm is the interface for the Hand's future form and, in this regard, constitutes its functionality in a pro-active constellation that not only adapts to the environment but also self-actively searches for appropriate grasping modes. Hence, there is an advanced level of contingent collective agency that may even be attributed a kind of intentionality (cf. Rammert 2008) that a new form needs to support.

However, the constellation not only imposes the range of possible forms. It also opens up *opportunities for new forms* by re-distributing agency. Depending on the perspective, the Meka robotic arm not only imposes the specific formats of robotic hands, but is also open to new forms. Its kinematic architecture has enough degrees of freedom to allow limitation of the Hand's movements to bending its fingers. In that sense, the arm allows simplicity and limitation of the

---

36 Rammert clarifies that the first two levels of causality and contingency are substantial definitions that can be delineated not only in human action but also in the behavior of technologies. In contrast, intentionality is an attribution using anthropomorphic semantics to capture the behavior of "smart" technologies (Rammert 2008,11).

range of required movements in the Hand.[37] Furthermore, the arm couples the causal agency of materiality with the contingency of algorithms, which allows varying agency levels of the ensemble's elements. Coupling digital programming and material efficacy is also a capacity of Arduino boards used in Baecker's studio. However, Arduino boards allow different enactments than the Meka arm. Due to their open design, flexible connectors, and micro controllers, they are tools for fast prototyping, or they become part of the form when integrated as an element of a stabilized relation. In this sense, their position can change from tool to form.

Significant in both constellations are, in particular, those entities that allow *connecting heterogeneous materialities* like analogue and digital technologies. In this sense, algorithms conduct the material and sign-based interfaces that need to incorporate a new form. The RBO Hand needs a proper algorithmic pendant in the steering software of the Meka arm. This is not trivial, but requires inquiries into how physical elements and algorithmic parameters are related. Digital and material realms are also placed in relation through the Arduino board, which substitutes Baecker's body and continues the movement of a "wafting area" in a new setting. Algorithms alter agencies from linear causality to contingent interaction. They do not animate any structure; rather, their agency is enacted through aligning algorithms and materialities.

Nevertheless, the empirical description shows how unstable constellations are. The infrastructure in the laboratory is hardly reliable, and the means allocated in the studio function in a good-enough manner. Deimel and Baecker enact their human agency for inquiring into the hybrid connections: their advanced technical knowledge enables them to program algorithms; their actions do not follow predefined paths, but situatively adapt over the course of interaction, and they follow specific intentions, such as contributing to a research problem or creating an aesthetic artifact. Hence, their actions entail causal efficacy, contingency, and intentionality that enables them to solve problems and anticipate how a constellation works before it is realized (cf. Rammert 2008, 10-1). However, the video sequences show how this assumed superior status of human agency is enacted through the dirty work of *bodily interactions*. Their bodies make them relate to the materialities that give form to a new object. In this regard, causal efficacy is not a low level of agency, but constitutes higher levels of contingent and intentional acting. This is demonstrated in both video sequences. In the first sequence, Deimel uses his body as a kind of proxy-resistance as he pushes the arm in order to see what the system state is. The second sequence illustrates how Baecker uses his body to anticipate whether a new connection is feasible or not.

---

37 In contrast to robotic hands with additional limbs, such as the Hitachi Hand from 1984 (Rosheim 1994: 204).

Deimel and Baecker enact their bodies to properly coordinate their activity with the malfunctioning technology and use their bodily senses to experience the relation of elements when those elements are coupled in a new way. In those situations, their bodies become epistemic tools for inquiring into the appropriate range and potential means for new forms.

This account stresses the agencies *through and in* which new forms evolve. On the one hand, relations between entities impose the range of forms, as specific formats are pre-scribed into them, but, on the other hand, hybrid constellations also enable new forms, as technological infrastructures are open to distributing agency among different entities. Nevertheless, the situated distribution of agency also shows how the agential location of a form has to be searched for, found, and fitted. This is how the interfaces to which a form connects are adapted and how entities are arranged according to the functionality that they are to take within or in connection to a new form. However, this account makes questions concerning novelty more difficult to answer. Novelty is both individuating and relating, but how does a form individuate in hybrid constellations that stress distributed agency? How does one account for the efficacy and agency of a new form that evolves within given and pre-structured ensembles?

3.3 Assembling Technical Forms

In the following and subsequent sections, I approach these questions and address the assemblage of the RBO Hand and *Mirage*'s forms and, from there, their relations to the hybrid constellations of their existence. A major challenge in addressing the assemblage of technical forms is finding a comparative structure. The structure needs to allow for a close comparison of technical features and capture how similarly or differently they are embodied, while simultaneously acknowledging the very different morphologies of the RBO Hand and *Mirage*. In this regard, I bring along the structure that I have used to sketch the technical formats and apply it in the following for the comparison of both technical forms with slightly adapted subheadings. Hence, the following comparison follows subheadings, such as embodiment, kinematic architectures, actuation, sensors, etc., that are abstract enough to capture technical relations in both forms.

For the sake of comprehending the technical descriptions, I place two figures of the RBO Hand and *Mirage* at the beginning of this section (*Figure 16* and *Figure 17*). The pictures capture both objects' material states, which marks the preliminary stabilization of the technical convergence described in the following. The figures include annotations of both form's technical components, whose assemblage is the focus of this section.

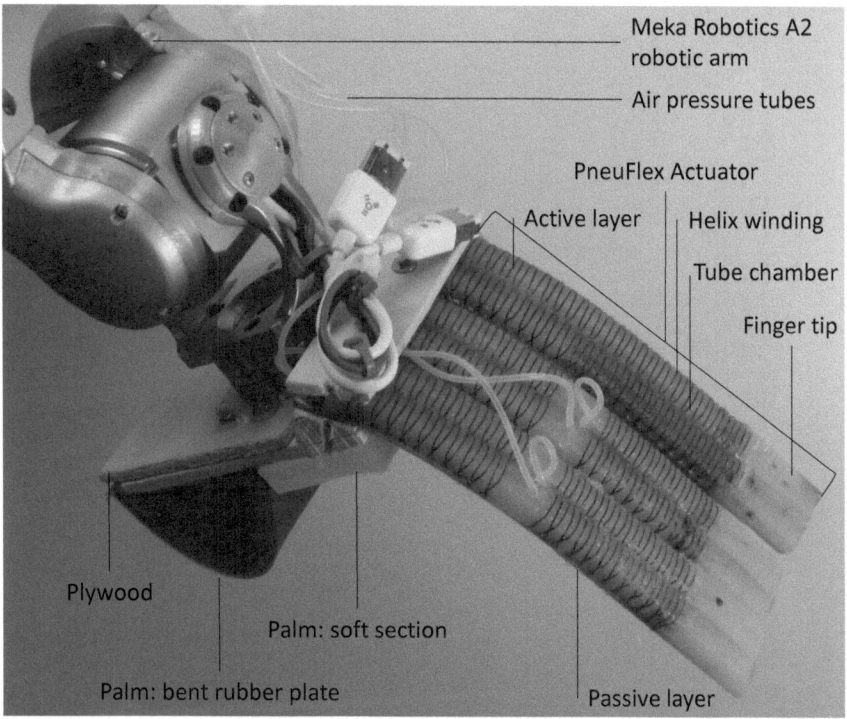

Figure 16: The RBO Hand's technical form (own annotations, Deimel and Brock 2013, Figure 1).

3.3.1 Embodiment as Epistemic and Artistic Stance

Above, I stress that the embodiment of concepts has an epistemic function in cybernetic machines. Ross Ashby, for instance, rejected simulating his *Homeostat* in favor of building his own "adaptive brain." As I outline in Chapter 2, the RBO Hand and *Mirage* are both objects with strong conceptual ambitions, and the stories of their origins embed them in larger discourses and differentiate them from the usual kinds of objects in the field. I stress that there are different articulations of ideas that make both objects different and drive the intent of their construction. These articulations already indicate that conceptualizing and manufacturing are not separate processes but are interrelated, because difference is enacted through heterogeneous entities such as stories, tropes, material traces, prototypes, and bodies – and not simply through a model and a finished material object. The interrelation of

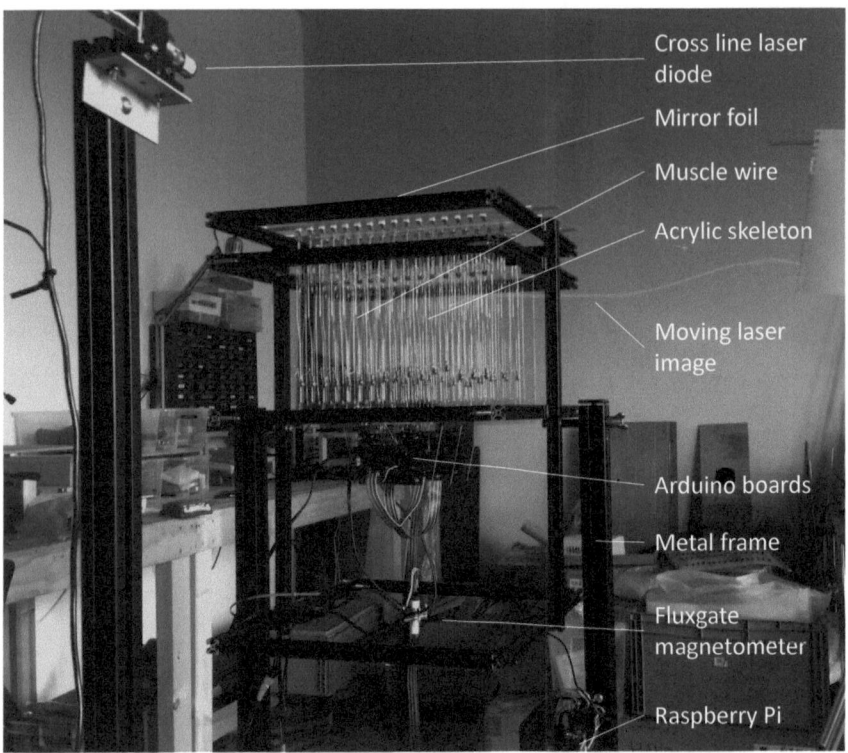

Figure 17: *Mirage's* technical form in Baecker's studio, close to its first exhibition (own picture).

conceptualizing and manufacturing stresses beginning the following description by asking what embodiment actually means for practice.

In general, robotics is a discipline in between science and engineering. Still, the relation between formalized theory and the dirty work of building technologies is ambivalent in robotics. When I spoke with the director of the robotics institute, Brock, he stressed this ambivalence and laid down his view on progress in robotics and the role of theory building. When asked about the role of mathematical formalization, which is integral to the scientific culture of robotics, he answered with a figure. He said that, for him, robotics research is comparable to the practice of "alchemists." Alchemists used to throw together all kinds of ingredients in hopes of gold. Although, nowadays, their practice has a much more mystical than scientific appeal, they still laid important stepping-stones for modern chemistry. This story entails Brock's stance concerning the relation of mathematical formalization and scientific practice. He is not against mathematics, he emphasized, but is for it only when appropriate for the problem under investigation; for him, robotics is still far from having identified problems that allow formalization. In his view, robotics is still in a phase of alchemy rather than having established scientific laws for building robots. He believes robotics makes better progress when procedures are explorative rather than when formalization is postulated. Brock said, for progress, one needs to ask how to create behavior, and only when the behavior created is understood is formalization an appropriate tool. In that sense, building the objects one seeks to investigate and testing an idea in its material form is an epistemic approach that is significant to the work in his institute and not given per se.

Brock's account allows for consideration of the RBO Hand as an embodiment whose material form is significant to the epistemic practice he regards as appropriate. This is a considerably different account of embodiment than building objects in order to legitimize one's research. Nevertheless, Brock also stressed that building a new kind of hand is a matter of showing the grasping community that unconventional and technically simple approaches lead to satisfying results when it comes to concrete grasping performance. However, this is not the point I want to make here. Rather, I want to emphasize what resonates in Brock's first account. *Brock considers embodiment as way to create behavior, which, in return, is what he seeks to investigate.* This account enacts embodiment as constituting progress. The step makes theories and concepts possible, not the other way around.

For *Mirage*, embodiment has a similar meaning, although it does not attempt to advance shared theoretical knowledge. As I have already mentioned, for *Mirage*, Baecker attempts to give body to the abstract concepts of learning algorithms. He refuses to continue his hallucinating machine as an algorithm, rendered visible through a computer animation or the like. In that sense, there is no enforced requirement of a material body so as to visualize how unsupervised learning produces contingent patterns. Instead of going the presumably easier route of designing a graphical representation of hidden activities, he turns to a material embodiment of his ideas. In the context of his artistic practice, this is more than merely choosing a format. Rather, embodiment is what produces the aesthetics of his artwork. Similarly to historical cybernetic machines, the agencies of different materialities and their complex and contingent interplay produce movement and a visual image. Hence, in Baecker's approach, embodiment cannot be considered as representing concepts, but foremost as creating technical behavior as an aesthetic performance of concepts.

3.3.2 Materials and Manufacturing Method

As indicated above, there is a vivid discussion over the choice of materials in robotics. The choice of materials affects most of the features of a robotic hand, such as its weight, compliance, and strength. Furthermore, the choice of materials also influences to what scientific domain a robotic device may contribute. The domain of soft robotics, for instance, is rather young and open to various applications, as long as they exploit the adaptive capacities of soft materials. In his story about the origin for the idea of the RBO Hand, Brock referred to both aspects. Based on his experience in previous experiments, he searched for a way to build a competent hand and found approaches to doing so in the emerging field of soft robotics. The material characteristics of silicone opened up opportunities for realizing this particular kind of grasping, which is less controlled by algorithmic planning than by the hand's autonomy. When asked about the main advantages of using silicone, Deimel answered that it is very flexible and expansible up to ten times its deflated size. That is why complex and large deformations are possible without complex joints. Deimel stressed that the material's compliance compensates for the imprecise positions of the hand and, hence, allows for robust grasping without exact information about an object or hand's location. Silicone is a material that is not very precise, but it is also very fault-tolerant, stated Deimel.

Furthermore, silicone is a material that is easy to work with, as its processing does not require tremendous investments or exclusive *a priori* knowledge. Hence, it allows for experimentation and tinkering. In order to exploit the material's capacities, one can, for instance easily change the shape and arrangement of the fingers or vary the material's softness by altering the ratio between two rubber components. As reported, Brock was in a position in which he knew a colleague who already had experience in working with silicone and, hence, was able to tell him what tools he needed to work with it. After buying the main tools, such as a vacuum pump and chamber, he hired Deimel. Deimel not only became the person who does most of the research with the RBO Hand, but he is also the one who builds it. It is important to note that Deimel did not have any experience in working with silicone or in manufacturing robotic hands. He is educated in cognitive science and informatics. Nevertheless, he was able to work with the material and its manufacturing method easily from the beginning and was already able to do experiments with the first prototypes within his first year.

Not only is the simplicity of working with silicone effective for Deimel's immediate work, he also wrote a tutorial about the manufacturing process that can be accessed through the institute's website.[38] The tutorial does not consider the complete RBO Hand, only its silicone fingers – the PneuFlex Actuators. A less detailed version of the procedure is part of the first published research paper on the RBO hand (Deimel and Brock 2013, 2042). According to both sources, the manufacturing process for the hand's silicone fingers can be summarized as follows (cf. *Figure 18*):

1. *Active layer casting*: The active layer is the main silicone body of each finger. Each finger is approximately 9mm wide and thick, and 130mm long. The mold for each finger is 3D-printed. The silicone mold features longitudinal chambers for the tubes, which are later inflated with air. They are shown in the fingers' cross-section and longitudinal cut in *Figure 18*. Furthermore, the mold has small ridges that imprint groves along the edge of the silicone in order to hold the thread of the reinforcement helix in place. The casting requires evacuation of the mold in a vacuum chamber and the filling in of the silicone.

2. *Reinforcement helix winding*: This step is the winding that keeps the silicone in form when inflated. It consists of common polyester sewing thread, size 50. The turns are approximately 4mm apart in order to distribute the strain evenly throughout the silicone. To neutralize torsional force, there are two helix windings in opposing directions.

3. *Passive layer casting*: The passive layer is each finger's bottom layer, which is colored blue in *Figure 18*. The main manufacturing difficulty is not in handling the material as such but in gluing the active and the denser passive layers together. Hence, the passive layer is reinforced with a porous fabric embedded into and permeated by silicone. This allows robust bonding and enables the transmission of forces. The passive layer is approximately 1mm thick after degassing in a vacuum chamber.

4. *Assembly*: The active layer is placed on top of the passive layer before the silicone sets. In the final step, the embedded air chambers are connected via silicone tubes (0.5mm inner/1.5mm outer diameter), which are inserted into the silicone using a 2mm cannula and sealed using pasty silicone adhesive.

These four steps form the material basis of the RBO Hand's design. The detailed documentation and publication of the manufacturing process indicates that this is not merely pragmatic but also encouraged by showing off the simplicity of its design and enacting the Hand's easy manufacturing as a trademark. Remarkably, these activities focus on the finger's design and the use of silicone as an unconventional, new material. Everything else that is required to make the fingers become a robotic hand is not necessarily black boxed, but surely not paid much attention to in the tutorial or research paper.

---

38 Website of the tutorial by Deimel: http://www.robotics.tu-berlin.de/menue/research/compliant_manipulators/pneuflex_tutorial/ (last accessed March 18, 2015)

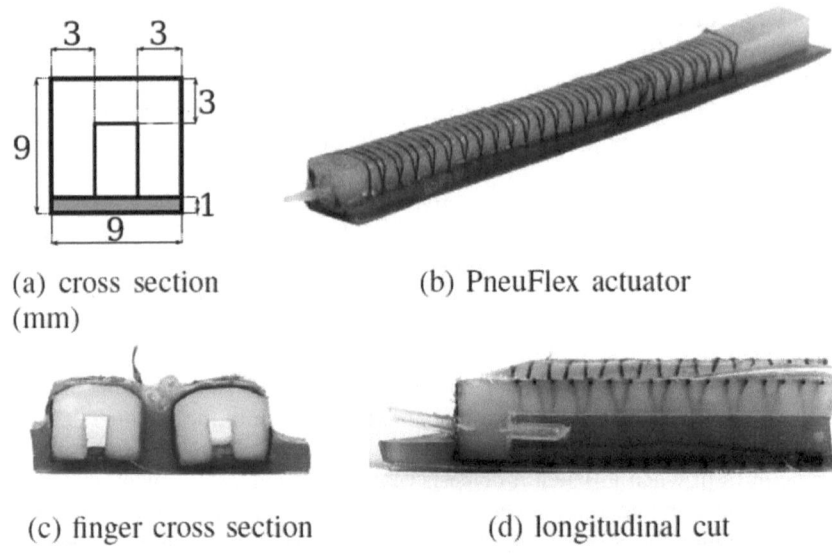

Figure 18: The composition of the PneuFlex Actuator (source, Deimel and Brock 2013, 2040, Figure 2).

A major difference concerning the materials used for the RBO Hand's and *Mirage*'s designs is that the silicone was a stable element in the Hand's genesis, no matter what form it took throughout my two year ethnography, as opposed to *Mirage*, whose materialities changed over the course of its design. *Mirage*'s materials went from wooden plates, strings, and elastics to acrylic glass, laser light, and a metal frame. None of the initial materials stabilized in the time between the articulation of ideas and the exhibited installation.

However, despite this difference in the object's material compositions, intentions to choose a specific material were quite similar. For instance, Baecker stores several wooden plates in his studio, as rapid prototyping requires a material that is easy to process. Similarly to silicone, materials such as wood, strings, and mechanical elements do not require exclusive knowledge in order to manufacture quick prototypes; they are easy to change in size and shape and to connect with other elements. Furthermore, they do not require sophisticated tools or knowledge to be processed, and they are available in regular hardware stores.

However, as *Mirage*'s technical form advanced, easy processing became less significant to the choice of materials. The use of acrylic glass may serve as an example of this shift. At a basic level, it allows for rapid prototyping, as de-

3.3 Assembling Technical Forms

scribed in Section 2.1.2. Plates of acrylic glass were at hand in Baecker's studio, since he had used them in previous installations. They were easy to cut and could serve as frames for moving elements. In addition, the material is not only pragmatic, but also carries the specific semiotics of Baecker's art. He stated that acrylic glass is a very clean material that somehow transports the aesthetics of scientific experiments. This is one reason why he had chosen to work with it before. Nevertheless, this choice is bound to specific conditions. In order to integrate acrylic glass into the design of an installation, not only using it for prototyping, he needed precise elements that were custom-made by a laser-cutting specialist. Such cutting is rather expensive and only worth the effort if he gets an exact shape that he can use for the finished installation as well as an exact number of elements. These conditions influence what material he chooses and, furthermore, how the material stabilizes the form of the object. It is only worth investing in custom-made acrylic elements if he is sure enough that their shape are no longer going to change and that they will make it into the finished piece. This is how acrylic glass is at once a material at hand that allows for rapid prototyping and, on another occasion, a material that requires anticipation of the finished form.

This acrylic glass example also illustrates how pragmatic and aesthetic intentions mingle in the choice of materials. In that sense, all other materials used for *Mirage* were chosen according to both aspects. For instance, the metal frame, is on the one hand, appropriate because it is heavy, stable, and to be cut with a regular saw, and, on the other hand, it is a clean material associated with tools and machines, which brings about an aesthetic tension in contrast to *Mirage*'s fluently moving image.

In Chapter 2, the moving image was embodied through strings, wires, and Baecker's body. For the finished installation, the wires remain as means of actuation, but the moving image is generated by a laser projection. Differently from strings and wood, the laser is a technology that requires advanced knowledge of electronics. It is voltage- and ampere-sensitive and requires alignment of the strength of the diode with the connected technical conditions. Baecker reported that he had thought about using a laser for a long time, but only got to work with it through a colleague at the time when he was already in the process of designing *Mirage*. Hence, he did not have access to laser technology at the time of my first ethnographic visits in the studio. When I visited him in his studio, after he had chosen to experiment with a laser, he had prepared a test setup for which he tightened a laser diode in a vice[39] and focused the light beam onto a small mirror (*Figure 19*). The laser was a cross-line laser, which does not produce a focused

---

39 A metal tool with movable jaws that are used to hold an object firmly in place while work is done on it (Oxford Dictionary).

point like a laser pointer, but a horizontal line whose measurements depend on the angle the diode allows and the distance from the projection surface. The mirror is bent so as to break and redirect the laser light onto the wall. A crooked image appears on the studio wall, whose turbulent line is determined by the mirror's bends and buckling. Baecker told me that he likes the aesthetics of laser light, as it has an almost haptic appearance.

Choosing a laser as a material changed *Mirage*'s technical and aesthetic genesis drastically. By entering laser technology into the design, *Mirage* became an installation based on dynamics caused by light reflections. Unlike string, a laser needs a source (the diode) and a projection surface to become visible. Making a laser move is only possible via reflections caused by a moving surface. In that sense, *Mirage* requires several sub-ensembles (cf. Simondon [1958] 2012a) to enact a moving image, as the laser diode is a technical unit and the moving surface is another unit that works separately.

The choice of materials for the RBO Hand and *Mirage* share distinct similarities. One similarity is in regard to the knowledge needed to process materials. For rapid prototyping and tinkering practices, easy handling of materials is an advantage of silicone, wood, metal, and string. In order to start working with new materials, Brock, Deimel, and Baecker drew on external people, such as research colleagues or artists, who provided information on specific tools and handling practices. This more exclusive information based on other people's experiences is significant for engaging with previously unknown material like laser technology.

Another similarity regards the material's significance to making both objects distinct. The description of the RBO Hand already indicates how difficult it is to focus on the pragmatic use of silicone alone. Silicone is both a pragmatic technology that contributes to solving the problem of grasping and a discursive statement that makes the Hand distinctively different from others. This is similar to the laser used for *Mirage*. A laser as such might not be a discursive object in the media art field, but, for *Mirage*, it becomes a visible trademark. Using laser technology is a choice of how to embody the idea of a hallucinating machine, and, furthermore, it entails that *Mirage* requires darkness for its exhibition. This is not arbitrary, but a choice made in the design, which transduces to situations that enact *Mirage* as a distinct piece of art.

3.3 Assembling Technical Forms 139

Figure 19: Early experimental setup of cross-line laser and mirror (own picture).

3.3.3 Kinematic Architectures

In terms of robotic hands, the kinematic architecture articulates the choice of degrees of freedom and the numbers of joints and actuators to realize the movements. In order to compare the RBO Hand and *Mirage*, I regard the kinematic architecture of a form in more general terms as the schematic of what and how technical relations are implemented in an object.

In their research paper, Deimel and Brock have given an account of the RBO Hand's general parts and technical structure (Deimel and Brock 2013, 2042-3). They considered the Hand as a prototype. The authors wrote that they chose to give the Hand three fingers, each consisting of two parallel PneuFlex Actuators. The total length of each finger is 130mm. One of the three fingers has two actuators of 100mm length each, with an unactuated tip of 30mm length. The other two fingers are divided into two sections, which improves their grasping performance. The sections are 35mm and 45mm long and separated by a 10mm passive piece. Each finger has a cross section of 9×21mm. Two supply tubes, one for the actuators on the outer part of the split fingers and one for all the other actuators, control the finger movement. The fingers are mounted on the "metacarpal" plate at a 30° angle.

The palm is split into two sections: a flat pad of very soft silicone (translucent, 10mm thick) and a bent rubber plate (blue tinted, 3mm thick). The latter structure creates a soft pad to match item shapes. The frame of the RBO Hand consists of a plywood plate (80x80x3mm), which holds the silicone parts and, furthermore, allows for the Hand to be mounted on a robotic arm. In their paper, Deimel and Brock highlighted that they found this assembly to be robust, easy to manufacture, and able to be adapted quickly during rapid prototyping. An external reservoir supplies pressurized air to control inflation and deflation of the hand.

Based on this architecture, the typical kinematic ratio between degrees of freedom and joints/actuators is difficult to express in numbers. For instance, the Salisbury Hand consisted of nine degrees of freedom and nine joints. In contrast, the RBO Hand does not define its degrees of freedom, as its grasping concept is based on interaction between the given item and the Hand. The item's shape determines how the Hand deforms. Hence, it makes little sense to explicate how many possible degrees of freedom the Hand has. Rather, pictures demonstrate deformations and show the Hand grasping items of different shapes. In order to trigger grasping movements, only one signal is required: inflating/deflating. The Hand's architecture allows two of these binary signals to be operated via its two supply tubes. In sum, this makes the Hand a highly underactuated robotic hand.

Mirage's architecture has undergone significant changes compared to the test structures analyzed earlier. Whereas these structures were either a net in an acrylic glass frame or a net of strings that was spun between two plates, for its exhibition, *Mirage* is changed into an installation that has an outer metal frame of approximately 150x50x50cm (height, length, width) and generates a laser image.

One reason for the metal frame architecture is the laser. Whereas Baecker used a vice for his first laser experiments, he anticipated that he would need a large stand to hold the laser diode and allow it to point downwards for the finished installation. That is the reason for the metal frame's large post, which was able to hold the shaft of a laser diode. On the same base, but separated from the post, was an inner frame for a mirror. Later, that inner frame held the acrylic elements and muscle wires, which I report on below. The inner frame is flexible so as to allow for vertical alignments of the laser light and reflecting mirror.

Another reason for the metal frame is mentioned above: *Mirage* is designed as a standalone piece. Unlike the RBO Hand that is designed to connect to the Meka arm, which is made possible through its plywood plate, *Mirage*'s form is not influenced by such infrastructural requirements. It is supposed to stand alone like a sculpture and work with no more technical requirements than electricity. The frame is a means to attach *Mirage*'s technical components and set the boundaries for their arrangement. In that sense, the frame captures the allocation of components and renders the outer appearance of a standalone apparatus.

Nevertheless, the metal frame captures but does not determine *Mirage*'s kinematics. Comparing the kinematics of *Mirage* with those of robotic hands is ambivalent in any case, as *Mirage* is not described in quantities like degrees of freedom or number of joints. Still, *Mirage* has kinematics that can be described in their allocation and efficacy. Comparable to the degrees of freedom of a robotic hand, *Mirage*'s kinematic architecture determines the range of movements of the laser projection. The range of these movements can be expressed in directions, upward and downward, and size, depending on the distance from a wall. Furthermore, the inner complexity of the image, or its bending and buckling, is caused by actuators that are quantifiable. At its finished state, 48 wires move *Mirage*'s image, arranged on three acrylic skeletons. These skeletons are attached to the inner metal frame. For the sake of quantification, this makes one degree of freedom (upward and downward movement) and 48 joints. Roboticists would call this a highly overactuated technical structure.

3.3.4 Actuation

The RBO Hand's actuation principle is already described in its material structure. The PneuFlex Actuators are driven pneumatically through the inflating and deflating of the fingers' inner tubes. For experiments, the fingers are driven with an air pressure of 210 kilo Pascal, which produces a force of 1.5 Newton (Deimel and Brock 2013, 2043). As indicated in the description of their material structure, the PneuFlex Actuators consist of two separated tubes in order to optimize the Hand's grasping performance. Nevertheless, the Hand's movements are simplistic, as it is only one signal that triggers a grasp. This entails that the Hand is always driven in the same way and in full capacity if commanded to grasp. Nevertheless, Deimel stressed in an interview that the softness of the silicone allows complex deformation based on such simple signal processing. Hence, the research paper expresses the Hand's versatility not in technical parameters, but demonstrates it through pictures that show how the Hand successfully grasps items with very different surfaces. Such items range from tubes, water bottles, and cylinders up to more complex surfaces like those of sunglasses or a tape dispenser, which are difficult for robotic hands to grasp.

Movement is central to Baecker's idea for *Mirage*. As described in Section 2.1.1, he thought of a hallucinating machine and a wafting area. He enacted such images in test structures made from strings and elastic elements, and through his body by making a snake movement with his arm. The strings, which he used in the test structure in Section 2.2.3, were means of actuation and, at the same time, materialized the image. For a previous installation of his, he made use of that same concept. His installation *Rechnender Raum (Computing Space)* consists of a wooden frame that holds a complex mesh of strings that is actuated by mechanical pulleys and digitally steered motors – here, the image is materialized through the means that equally transmits actuation.

Introducing laser technology to the design of *Mirage* has lapsed this concept. Laser light is a material very unlike strings. It has no haptic resistance; it cannot transmit actuation and requires an external projection surface. These characteristics urged Baecker to inquire into means of manipulating how the light is projected and how to make the projected image move. A test setup for this inquiry into manipulating the projected light is already mentioned above. Baecker's first laser experiments inquired into how different mirror shapes influence the projected light. Already in the early stages of experimenting with the new material, Baecker stressed that it was difficult to figure out an appropriate way to make the image move. He did not want the light source to move like in popular laser shows, but to enact movement by breaking the light beam. This creates a more detached and contingent movement and should make the image appear as if it is floating, said Baecker. Hence, the reflecting surface or mirror needs to move in order to create a moving image on the wall.

But before Baecker chose to use mirror foil to reflect the laser, he tried silicone oil as a surface for reflection. Silicone oils are available at high viscosities and move slowly and are resistive when swirled in a basin. Hence, they have characteristics similar to the wafting area Baecker had in mind. But he discontinued use of such liquids, as they were too messy to work with and did not solve the problem of how to actuate movement. Hence, he turned to a mirror foil that was light and flexible and, hence, could be moved with less force than silicone oil in a basin. Furthermore, foil requires only small trigger movements in order to manipulate a laser projection. That is, if a foil is pulled down about 2mm at some point, this small diffraction translates into a rather large bending of the projected image. This allows low actuation to translate into large aesthetic effects.

3.3 Assembling Technical Forms 143

So few requirements in terms of distance could be implemented through several actuation principles. When I visited Baecker during some of his first tests, he said he was looking for a maximum movement of about 5mm. For *Rechnender Raum*, he used electronic motors that actuated a varying string movement of about 2-4cm, which then translated into other mechanical pulls and larger movements, he reported. However, using electronic motors was not appropriate for moving foil, as he sought to pull down the foil not only at one or two points, but at several densely allocated points. Hence, he had little space for actuators, which made using motors like those used for *Rechnender Raum* unsuitable. Nevertheless, the decision for foil, as well as for many points of actuation instead of only one or two, allowed Baecker to express his problem more precisely and seek a solution. In his studio, Baecker showed me a small structure of acrylic glass that consisted of six bows arranged in a row on a small platform, which I have already mentioned earlier (cf. *Figure 15*). He showed me how he had spun a wire between the ends of each bow. This was muscle wire, he explained. Muscle wire is an electronic wire that contracts when fed with an electronic signal and returns to its original position after release. It is also sensitive to different signal strengths. Muscle wire belongs to a larger class of shape-memory alloys and is used in robotic hands, too.[40] In robotic hands, it is implemented as tendons to transmit actuation. For *Mirage*, muscle wire possesses the advantage that actuation and actuation-transmission are realized in one element and, hence, render other motors and strings obsolete. Baecker showed me the small test setting, for which he attached the six wires in a bow to six analogue outputs of an Arduino board. By programming the output voltage via a micro controller, Baecker was able to test what signal strength caused what contraction. He showed me how each output was able to produce an individual signal that made all six wires contract in a sequential pattern. This contraction made the bow bend, as the acrylic glass was flexible enough to adapt to the stress. These tests took place approximately six months before the exhibition of *Mirage*. From then on, muscle wire has been the stable means for actuation. Nevertheless, for the finished installation, Baecker advanced the rudimentary assembled test structure toward three skeletons made of acrylic glass, which were laser-cut especially for the purpose of *Mirage* (*Figure 20*). One skeleton can hold 16 densely arranged wires. Each wire pulls down a spring that is attached to another acrylic hook. Hence, the inner flex of an acrylic bow is no longer exploited, but an additional spring and hook transmit the actuation of each muscle wire.

---

40   Shape-memory alloy was a trademark material of the Hitachi Hand when the robotic hand was introduced in 1984 (Rosheim 1994).

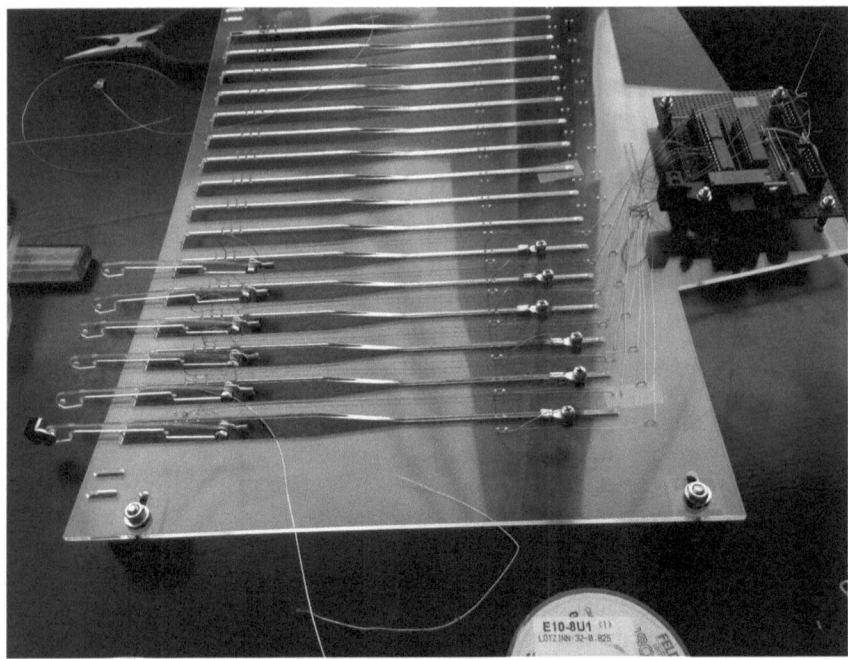

Figure 20: Skeleton made of acrylic glass (own picture).

3.3.5 Sensors and Input Signals

Sensors are a pivotal issue for the RBO Hand and are stressed in research papers as a significant difference in the RBO Hand when compared to common robotic hands. The RBO Hand's difference is not in a special kind of sensor, but in that it does not have any sensors at all. Hence, the hand's grasping style is not based on planning, but on compliance. Planning requires that a robot is equipped with sensors that a) obtain visual information about an item that is supposed to be grasped in order to detect its morphology and b) sense what characteristics the item has in order to control applied force, that is, if it is soft, solid, fragile, etc., (Controzzi, Cipriani, and Carrozza 2014, 234-5). Deimel explained to me that, in contrast, the RBO Hand is explicitly not based on planning that requires sensors, but on the interaction for which the compliant characteristics of the Hand's material are exploited. That is, there is no information sensed before the Hand performs a grasp.

However, there are different enactments of this technical difference. Firstly, when Deimel or Brock explain the approach behind the RBO Hand, they refer to the lack of sensors as an exploratory difference that makes the Hand's design simpler than typical designs but with compatible grasping performance. Simplicity is commonly regarded as good, because it makes designs robust and less prone to failure. Furthermore, exploiting material characteristics concretizes the Hand's functionality, as it makes pluri-functional use of silicone, which not only is the Hand's body and force transmission, but also implements how the Hand adapts to its environment. However, secondly, in order to run publishable grasping experiments, the Meka arm needs to move the Hand in an optimized position, and doing so requires fixed *a priori* programming of arm movements, as the Hand is implemented without sensory feedback (Deimel and Brock 2013, 2043). Sensory feedback would require visual sensors in the Meka setup and a corresponding kinematic architecture, as well as force sensors on the contact surface. Hence, in their research papers, Deimel and Brock address the lack of sensors not only as a positive trademark, but also as an active research topic. In a later article, they state that it would be desirable to integrate strain and touch sensors, but this is a very difficult task (Deimel and Brock 2014). In this sense, quitting sensors is a technical achievement of the RBO Hand's distinctive design, but it may not be a desirable novelty in every situation.

In contrast to the RBO Hand, *Mirage* is equipped with a sensor: a fluxgate magnetometer[41] that senses changes in the Earth's magnetic field. The magnetometer is sensitive and registers frequent minimal changes in the field. These changes occur randomly without predictable patterns. Fluxgate magnetometers consist of a magnetically sensitive core wrapped by two wire coils.[42] One coil is charged with an alternating electrical current that causes an irregular cycle of magnetic saturation (i.e., magnetized, unmagnetized, inversely magnetized, unmagnetized, etc.). This constantly changing field induces a corresponding electrical current in the second coil. A detector measures their output current. Matching input and output currents indicate a magnetically neutral background. In contrast, differences in input and output indicate that the core is saturated in correspondence to a background magnetic field. How far both signals are out of step depends on the strength of that background magnetic field. The magnetometer used for *Mirage* transmits such differences as an analogue output voltage that is proportional to the magnetic field.

---

41 The fluxgate magnetometer was invented by Friedrich Förster (1908-1999) and, hence, is also called Förster-Sonde, as in some publications concerning *Mirage*.
42 The following paragraph is a description of how fluxgate magnetometers generally work. Its major source is the corresponding Wikipedia entry: http://en.wikipedia.org/wiki/Magnetometer#Fluxgate_magnetometer (last accessed: April 8, 2015).

The characteristics of the Earth's magnetic field as input signal is significant for *Mirage*'s design. Firstly, such signals are typical for cybernetic machines, as they are random and, hence, require an adaptive machine to register and convert them into a specific output. Given that an adaptive design is intended in the first place, one could also say random sensory inputs allow for sensible adaptive designs. Secondly, the signals render a force that is not perceived by humans. In opposition to robot sensors, they do not have an anthropological ideal like touch or visual signals, but make visible what is not perceivable via human senses without technologies. These two characteristics are significant because they correspond with the idea of a hallucinating machine; they are beyond control and render immaterial agencies.

Surprisingly, Baecker integrated the fluxgate magnetometer into *Mirage*'s design rather late. Baecker had worked with signals with similar characteristics before, such as, for instance, the electrical charge of minerals. Hence, it was not necessarily a major challenge for him to think of means to generate random signals that refer to a kind of hidden agency. Approximately three months prior to *Mirage*'s first exhibition, he told me that the metaphor of a hallucinating machine was his central concern throughout designing *Mirage* and less what environment he rendered through its image. This meant that he needed signals that were random but still changed rhythmically on larger time scales, like being asleep and awake. In principal, he said, he could have used any kind of signal source with such characteristics. For him, he said the magnetometer worked fine, as the Earth's magnetic field changes rhythmically depending on daytime or nighttime, while, on a micro scale, changes are contingent and appear randomly.

3.3.6 Converters

As the RBO Hand is (so far) designed without sensors, there are no signals that need to be converted for subsequent technical processes. Other robotic hands, whose movements are steered with sensory data, need such conversion through integrated micro controllers. However, the silicone material converts the resistance of an item into robust grasping without further technical support. There are no feedback loops, so to say. However, the Hand's movements are based on physical conversion but in the opposite direction. As said, no outside signals are converted, but signals are translated into movement from inside the ensemble. The air pressure that inflates the tubes of the PneuFlex actuators is converted into a grasping movement a) through the silicone's adaptive material capacities, which cause the actuators to deform, b) through the passive layer, which is glued onto the inflated active layer and makes the active layer form an encapsulating shape, and c) through the helix winding, which prevents the active layer from blowing up and instead converts air pressure into force. These processes are a conversion, but not in a cybernetic sense. The RBO Hand does not convert random environmental signals to regulate and adapt its behavior like a *Homeostat*. Rather, the described conversion, from the inside, is a designed and controlled technical relation that converts one type of energy (air pressure) into another (grasping force).

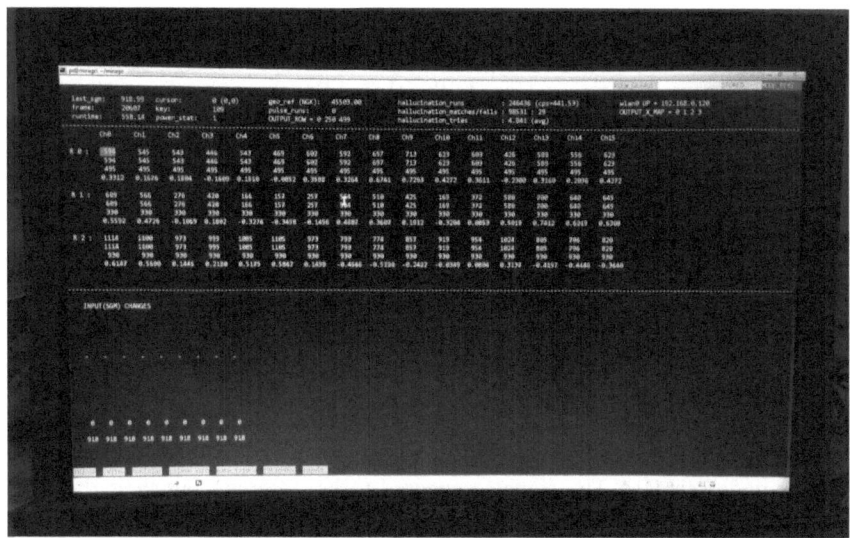

Figure 21: Computer screen showing frequently changing signals and their algorithmic variation (own picture).

In contrast, *Mirage* converts signals in a cybernetic sense. As said, the fluxgate magnetometer tracks random signals that constitute the machine's environment. Through its described functionality, the sensor converts the energy of a magnetic field into an electronic voltage that is sensitive to the field's strength. Here, the environment becomes an electronic pattern that feeds the machine. This is the first constitutive conversion in *Mirage*'s technical setup.

Nevertheless, this conversion does not take much design effort, since it is performed by the sensor, which forms a concrete technical unit. As Baecker said, it could have been any kind of signal with similar characteristics. Important to him is rather what happens afterwards, what embodies the hallucinating machine, and this is how signals are processed and converted and not so much the choice of signals that are rendered. Hence, the mechanism that converts the analogue electronic signals of the sensor is the central embodiment of his idea.

When I visited him in his studio, Baecker explained how he planned to process the signals from the magnetometer in order to actuate the muscle wires. During that visit, the setup to transmit the sensor signals was already in place. The analogue electronic signals coming from the magnetometer were connected to a Raspberry Pi.[43] A Raspberry Pi is an open hardware minicomputer comparable to an Arduino board but with extended functionalities. The signals fed into one of its analogue inputs. As the sensor was sensitive to minor changes in the Earth's magnetic field, the signals could be expressed in detail, spanning the whole range of three-digit numbers (in contrast, i.e., to a binary signal). We sat in front of a computer screen close to the sensor-Raspberry Pi setup. The screen showed a graph with three rows of signals that changed their positions frequently (*Figure 21*). I wondered why there were three rows, since only one signal was coming from the magnetometer. Baecker told me that he had logged the sensor data for 24 hours. These signals were running in the middle of the graph. In order to make things more complex and generate three rows of related but different signals, he had programmed an algorithm that produced an artificial signal. Those signals were being generated by a rule based on a random selection of patterns occurring in the logged data. Baecker explained that he would take, for example, a row of five signals and synthesize a new signal from these based on the probability of what signal would follow next. That way, he could generate a signal that appeared as a simulation of what was likely to happen next, but which was in fact only a variation of the original signal. He did so twice, so as to have three rows of signals. The signals appeared to be different, but their rhythmic changing vaguely indicated a relation. Baecker continued and said that this was how he understood his metaphor of a hallucinating machine.

3.3.7 Outputs

The Difference in Grasping Types

Generally, the RBO Hand's output is grasping. But grasping is not simply grasping an item in robotics research. Robotic hands are evaluated regarding their performance in grasping items, and, hence, there are theories about desirable grasps and means of distinguishing between types of grasping. Domenico Prattichizzo and Jeffrey C. Trinkle have regarded the most desirable robotic grasp as one that maintains control in the face of unknown disturbing forces (Prattichizzo and Trinkle 2008, 672). Grasp maintenance means that the contact forces applied

---

43 Website for the Raspberry Pi platform: https://www.raspberrypi.org/ (last accessed April 9, 2015).

by the hand prevent contact separation and unwanted contact sliding. Grasps that are maintained under every possible disturbance are called "closure grasps." A grasp with complete restraint prevents loss of contact and thus is very secure. For instance, the Salisbury Hand performs a closure grasp by closing its fingers around an item and pressing it against its palm. Prattichizzo and Trinkle point out two primary restraint properties:

(a) *Form closure*: A form closure grasp guarantees maintenance of contact as long as the links of the hand and the item's surface are well approximated and as long as the joint actuators are sufficiently strong. Form closure occurs when the palm and fingers wrap around the object forming a cage with no wriggle room. This kind of grasp is also called a "power grasp" or "enveloping grasp" (Prattichizzo and Trinkle 2008, 681).

(b) *Force closure*: Force closure grasps require fewer contact points to achieve closure than form closure and rely on contact friction. A grasp that holds a pen in its fingertips so as to write with it relies on force closure. Thus, force closure requires the ability of a robotic system to control internal forces.

The primary difference between form closure and force closure grasps is the latter's reliance on contact friction. Of course, it is also possible for a grasp to have partial form and force closure, indicating that only a subset of the possible movements are restrained. Both types can be modeled according to how well a robotic hand applies them to certain item shapes. These models may then define an ideal grasp under form or force closure, and may be applied for evaluation. Hence, they make grasps comparable.

In their second published research paper, Deimel and Brock explicitly address the modeling of the RBO Hand's grasping as difficult due to the nonlinearities of compliant grasping and the large number of degrees of freedom in the actuators (Deimel and Brock 2014). In an interview, Brock emphasized that this causes difficulties regarding how colleagues perceive the Hand. However, for him, performance is a matter of what a hand can do and not what grasping models a hand realizes. The presentation of the grasping results of the RBO Hand somewhat reflects this opinion. All papers include several pictures of the Hand grasping different items, like paper cup, tape dispenser, or bottle. The pictures refer to the Hand's capacity to grasp heterogeneous shapes due to its compliant behavior. They show how one type of closing movement is sufficient to match a wide range of shapes, as the PneuFlex Actuators have intrinsic and unquantifiable degrees of freedom (Deimel and Brock 2013, 2043-4). In contrast, grasp models can be described in schematics.

Although the Hand is actuated in one direction only, there are two distinguished grasping types in the 2013 paper: sliding grasps and surface-constrained grasps. For the first type, the Hand grasps an item on a table from the side. The second type is a grasp for which the Hand approaches the item from above and makes use of the table in order to pick it up (*Figure 22*). The results of the paper show that the Hand performs both grasps with robust restraint. Whereas the first type is a rather simple kind of form closure, as it applies as much surface contact as possible to capture an item, the second type is a form/force closure mix, which is significant for the Hand's compliant grasping. The significance is in how the Hand makes use of the table in order pick the item up. In *Figure 22*, the Hand picks up a pair of sunglasses positioned on a table.

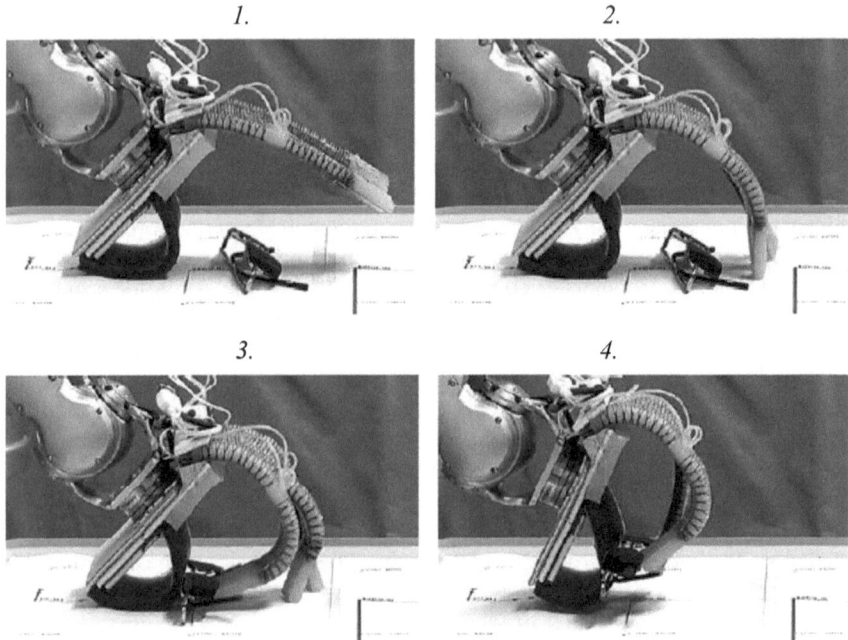

Figure 22:   The RBO Hand performing a surface-constrained grasp (source, Deimel and Brock 2013, 2045, Figure 9).

The Hand approaches from above and the grasp is described in four steps: 1. contacting the surface, 2. caging the object, 3. contacting the object, and 4. pitching to lift (Deimel and Brock 2013, 2045). This procedure exploits the Hand's compliance, as it makes contact with the table without information about its exact location, slides its fingertips over the surface, and pushes the sunglasses toward its palm. The Hand's softness makes the sliding movement possible as it adapts to the resistance of the table. Instead of sensing its exact position in relation to the table's surface, it simply slides over it without damaging itself, the table, or the sunglasses. Whereas sensor-based grasps regulate their grasping force through data feedback, the RBO Hand does not need this loop, since the silicone regulates the force of the air pressure by deforming its body according to the sunglasses' shape.

Existing grasping types do not capture well the described grasping style of the RBO Hand. Its restraint properties imply its environment, as it uses external surfaces to capture items. Robotic hands based on other designs have difficulties performing such grasps, as they require complex feedback data to adapt the hand's behavior. Furthermore, such grasping is difficult to compare, as it is not a property of the hand alone, but a capacity to interact with the environment. These interactions are somewhat beyond control, as the Hand's behavior cannot be modeled due to the nonlinearities of the silicone and its unconventional kinematics.

Image and Movement

Mirage's output is a moving image: it is the reflected light of the cross-line laser diode that is projected onto the mirror foil and from there toward the wall (*Figure 23*). The muscle wire actuates the foil, which makes the reflection move as the light adapts its beam to the continuously changing surface. The laser light is red and produces a focused reflection that does not lighten the dark room but remains at the wall.

Whereas I have discussed the RBO Hand's output in contrast to established grasping in robotics research, *Mirage*'s output does not articulate specific categories that correspond with shared evaluation criteria. Certainly, *Mirage*'s output is described in words, too, but figures like "synthesized landscape" are associative references to its aesthetics and not categories with which its output is evaluated. Hence, Baecker has commonly referred to his vision of the piece and not to field binaries as successful/unsuccessful to reflect upon his progress while tinkering.

Nevertheless, this does not mean that I regard *Mirage*'s output as the work of a genius whose objects are expressions of his mind alone. On the contrary, material conditions and technical relations structure Baecker's tinkering practices drastically. This is shown, for instance, in the previously described inquiry patterns. Nevertheless, contingencies caused by the complexity of technical relations also determine what *Mirage*'s output actually is. Conceptually speaking, *Mirage*'s output renders converted signals of the Earth's magnetic field as a moving laser projection; it renders hidden processes of interacting technical components. However, from observing *Mirage*'s genesis, one can tell that such narrative attributions mask the dirty work of concretizing technical relations. I want to draw out two examples of aligning *Mirage*'s features.

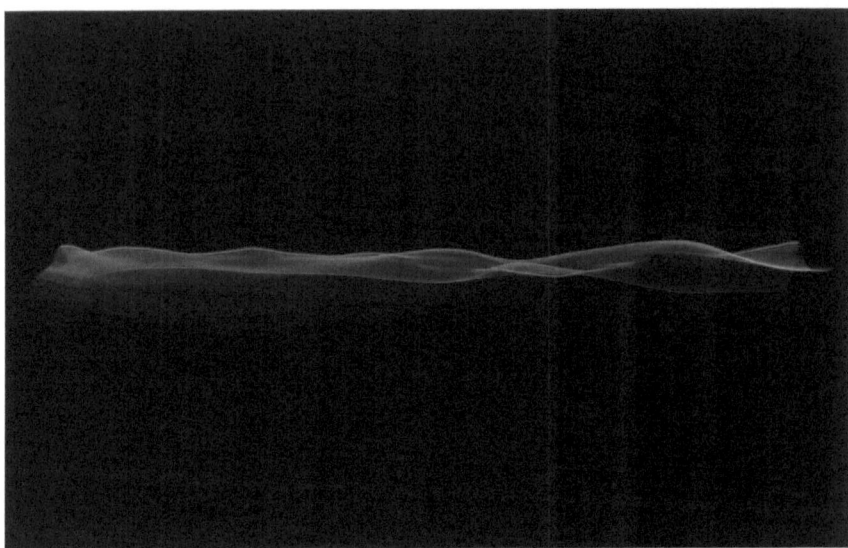

Figure 23: Mirage's moving image (source, Ralf Baecker).

First, connecting the acrylic hooks, which were attached to the muscle wire, with the mirror foil was a central concern of Baecker's close to the exhibition opening. For his first approach, Baecker used thin PVC foil with a non-transparent surface. The problem was that dints in the image occurred where the hooks were connected to the foil. Baecker used double-sided tape to connect the hooks and foil. The minimal dints that occurred where the tape met the foil translated into large and disturbing dints in the image. To solve that problem, Baecker placed small foam pieces between the foil and tape in order to decouple both elements and soften the connection. However, this was unsuccessful, as the tape's glue still left marks on the foil. Shortly prior exhibition, when *Mirage*'s architecture was almost finished, Baecker changed from PVC mirror foil to a foil made from acrylic glass. That foil was slightly thicker than the PVC foil and milky instead of opaque. Hence, the new foil was not a typical silver mirror with optimal reflecting capacities, but a foil whose surface only reflects laser light if it comes in at a flat angle. This trade-off concerning reflecting capacities was compensated for by its thicker structure, which remained bendable and did not pass on the small dints caused by the tape connectors. Substituting PVC mirror foil for acrylic foil was significant for considering the image as rendering the object's inner processes: although a surface that is less sensitive reflects the light, it was closer

3.3 Assembling Technical Forms 153

to the process it was to render, as the image's appearance was now structured through the actuated muscle wire and not through sticky tape.

Secondly, the preceding descriptions of *Mirage*'s technical features have already indicated that its output is not fixed but has to be calibrated on site. Hence, for its first exhibition, Baecker set up *Mirage* several days prior to the opening in order to fine-tune all elements. On site, he faced several problems. For instance, the grounding of *Mirage* did not work properly, which caused the failure of several microcontrollers. Furthermore, he realized that the image was sensitive to temperature. This entailed the problem that the image might occur differently depending on how many people are in the room or if it is daytime or nighttime. These site-specific variables forced Baecker to fine-tune the image in anticipation of conditions. There are four means to fine-tuning *Mirage*'s output on site:

1. The height of the laser diode changes the angle at which the light touches the moving surface. This influences the three-dimensionality of the image. If the light comes in steep, the image is only a horizontal line; if the light comes in flat, the image becomes three-dimensional.

2. The focus of the laser makes the image fuzzy or focused. It has to be aligned with the distance from the wall.

3. The inner frame is not fixed but can be skipped, which changes the height of the projected image on the wall. The change in its angle has to be aligned with the height of the diode.

4. The position of the double-sided tape, which connects the hooks of the muscle wire with the new acrylic foil, can be altered slightly. The more distant the three tape rows are apart, the more extensive the projected movement.

Baecker exploited these means to fine tune *Mirage*'s image until shortly before the opening. His objective was to keep the image stable under the changing conditions of the exhibition room; the image was to be altered by changing inner technical states and not due to the number of people in the room. Both examples of aligning *Mirage*'s output show how the complexity of technical relations and different materials structure what the image actually renders: does the image appear as intended through the algorithm that converts signals, or do redundant material properties and external conditions create the image's appearance? Aligning all the elements entails aesthetic, technical, and contextual considerations. *Mirage*'s image has to articulate all these aspects so as to be open to attributions that signify the image as a rendering of the machine's inner processes and not something else.

3.4 Articulating Novelty through Technical Forms

In the following, I attempt to connect the distributed agency of the hybrid constellation with the becoming of a new form. The perspective of distributed agency suggests the provocative question: who or what is actually grasping the item? A legitimate answer would be similar to Rammert's account of who flies an airplane. His reply is as follows (Rammert 2008, 16):

> "It is the hybrid constellation of people, machines, and programs. It is the mode how the agencies of the heterogeneous instances are distributed and connected with one another and the level of agency that is given to them in certain situations."

Similarly, the ensemble of the Meka arm, the RBO Hand, software, and Deimel, who maintains their relations, could be regarded as a collective hybrid actor that grasps an item. This is certainly a sensible answer, but only when it comes to stressing how technology works, and less how a technological object comes into being. Stressing that actions are distributed seems to contradict the regard for a technical object as new, as its functioning depends on several other technical units. Hence, the question to be addressed is what characterizes the functional relation between a new form and the hybrid constellation of its emergence.

Simonlon provided an entry point to answer this question. He stressed that technical genesis is characterized by *thresholds* to be crossed to enfold the potentials of an object's parts (Simondon 2012c, 5). Brian Massumi has explained that Simondon regarded such thresholds as the moments of invention; it is when "two sets of potentials click together, coupling into a single continuous system" (De Boever, Murray, and Roffe 2012, 25). Thresholds are the magnitude or intensity of energies and must be exceeded for a certain reaction to occur – they are the moments in which something comes into effect. This perspective shifts the moment of technical invention from a novel end that signifies a collective ensemble of people, machines, and programs toward a moment in the assemblage of an object's technical parts. The moment of invention is when a circular causality kicks in that unfolds the potentials of an object's components toward a new regime of functioning. This is like the spark of a combustion engine, which crosses the energetic threshold for the engine's parts to function as a self-maintaining system. Simondon stressed two aspects that concretize an object: the first is the internal resonance of elements, and the second is the pluri-functionality of an element, which means integrating redundant properties into a functional unit (Simondon [1958] 2012a, 19ff.).

However, I have already stressed that, despite of Simondon's technical examples, ideal types characterize his thinking; moreover, his examples, like the Guimbal turbine, are technical inventions whose inventive value is accounted for retrospectively as exceptional embodiments of technical progress. Hence, it is difficult to empirically search for moments of self-sustainability in the RBO Hand's or *Mirage*'s forms in order to account for whether they may be regarded as novel or not. However, Simondon's thinking can guide identifying moments or relations through which both forms' technical parts interlink and moments of circular causality kick in. In return, these moments are novel articulations of constellation and form.

In terms of the RBO Hand, the actors already address such kicking-in of synergies that unfold through the relation of technical parts. For instance, Deimel has stressed the non-linearity of silicone that makes modeling difficult, while Brock has stressed that creating behavior is more important to progress in robotics than new theories. Hence, both anticipate that there is a moment in the Hand's genesis where the form gains autonomy and exceeds their controlled modeling. In the assemblage of the Hand's parts, silicone is certainly a key component, and its design exploits the material's potentials. The manufacturing method manipulates the silicone in different ways. The form assembles an active and passive layer so as to exploit the silicone's deformability through the active layer, and the silicone's cohesiveness and friction through the passive layer. The Hand can only exploit these characteristics for grasping, because both layers are glued together as a unit that is bendable in a directed and controlled manner, supported by the helix twining around each finger. However, so far, these are only potentials, and the Starfish Grabber embodies them as well. What makes the RBO Hand different is its palm and the option to mount it on a robotic arm, which opens the object to become part of a technical structure. The palm is a passive part that supports the Hand's grasping as a resistant surface, against which the PneuFlex Actuators can push an item. This broadens the range of grasps compared to the Starfish Grabber, as the Hand may pull items over a surface and, from there, lift them. For this grasping, the Hand needs to be mountable onto a sophisticated robotic arm, as the arm moves it into an appropriate angle toward an item. As described in the Hand's output, this creates a distinctive style of grasping that performs grasps that are very difficult for other robotic hands to perform. For those grasps, the Hand exploits the silicone's compliance and integrates the material's deformability, cohesiveness, and friction, as well as the passiveness of the palm into a functional unit. This is the functional solidarity of components kicking in, which Simondon characterized as concretization, the pivotal moment of technical progress (Simondon [1958] 2012a, 19ff). Differently from Simondon, who would have regarded the technical infrastructure as imposing standardized forms, the Hand's unconventional design *builds upon* the abilities of the Meka arm for its new form. The point I want to make here is that the Hand's novel grasping style articulates the hybrid constellation. The technical infrastructure not only imposes standardized forms, but also allows new forms by distributing agency. The RBO Hand's form articulates novelty in the moment when heterogeneous entities connect: when the form combines the directed deformability of the silicone fingers with the passive resistance of the palm in such a way that both connect as a material unit, which is mounted on the Meka arm that drives it into a grasping position, from where the fingers can perform a grasp that exploits the material's compliance. This moment of connecting parts, when the hybrid ensemble performs a surface-constrained grasp, is the relevant threshold to be crossed and articulates novelty, because, from then on, the potentials of the PneuFlex Actuators become effective as a distinct type of grasping. The RBO Hand integrates parts as a functional unit whose synergies only unfold because they cross their potential's threshold by means of the hybrid constellation. The Hand becomes part of the constellation, and, through the clicking-in of components, the agency of the whole ensemble changes. *The form that concretizes the functional solidarity of its components, while equally building upon the abilities of the constellation to unfold its potentials, characterizes this articulation of novelty.*

For *Mirage*, the relation between the novelty of its form and the constellation of its emergence is different. Clearly, *Mirage* does not become part of a constellation that consists of other pro-active technologies like the Meka arm. The assemblage of *Mirage*'s technical features shows that the border between the technologies that are tools and conditions of its engineering and technologies integrated into its form is fluid. As said, the RBO Hand's form articulates the Meka arm's functionalities, but, for *Mirage*, the fluidity between the inside and outside of the form is material integration or rejection and does not script a stable set of materials. *Mirage* did not have a stable material core from the beginning like the RBO Hand's silicone; rather, different materials were tested, tinkered with, evaluated, and fitted into the evolving form or, respectively, excluded. A pivotal example of the fluidity of this boarder is the Arduino board, which Baecker used for testing just as he integrated it into the form – or the muscle wires that were at hand in the studio and later assembled with the acrylic bow as a unit that replaced body and strings as actuators. The selection of *Mirage*'s parts focused on their ability to connect to each other and the potentials that were unleashed through their connection. In some moments, *Mirage*'s assemblage concretizes its form in Simondon's sense; however, in other moments, it follows an opposite mode. A concretizing efficacy is found in, for instance, the integration of muscle wires, which are pluri-functional, as they connect two parts and work equally as actuators. In contrast, the integration of laser light made *Mirage*'s form less concrete than its preceding test structure of wooden plates and strings, because the strings are both actuators and image at the same time, whereas the laser diode is an additional unit with no other function than producing light. Similarly, the conversion from analogue sensor signals to digital algorithms and back to analogue outputs raises the number of technical parts and connections. The increasing complexity distributes *Mirage*'s agency among multiple and heterogeneous hardware and software entities. Through accumulative parts and the increasing number of connections, the level of agency increases in Rammert's sense: from causal efficacy to contingent behavior, which may be attributed a kind of intentionality, as the resulting light movements are directed according to opaque processes (cf. Rammert 2008). In *Mirage*'s form, the integrative accumulation of parts characterizes the relation between a new form and constellation. In contrast to the RBO Hand, whose constellation is open to incorporate a new form, *Mirage*'s form is open to incorporating agents of the constellation in which it evolves. Within this accumulation of components is a moment when *Mirage*'s form crosses a threshold that articulates its novelty. This includes: when the Arduino board converts the sensor signals; when the output signals trigger the muscle wires; when these pull down the springs that are double-taped to the acrylic foil; when all of the 3x16 contact points contingently make the foil move; when the laser light touches the moving foil; when the angle of the laser light is flat enough to create a three-dimensional focused projection. Why is this novelty? Because, when these parts work together, *the form articulates its complex technical connections and the aesthetics of a moving image*. This is when *Mirage*'s form individuates and articulates novelty.

The relation between form and constellation, which is different for both objects, shows that novelty is not a property of an object. Asking for form in the sense of Cassirer (Cassirer 1985) emphasizes the active character of finding connections and stabilizing them in a functioning set of relations whose elements become effective through their sequential order. This shifts the focus of defining technology from its end toward a definition through its becoming. The crossing of thresholds articulates novelty and makes parts function in a manner that exploits the capacities of their elements. This process individuates an object while still being relational, as it happens within a hybrid constellation that incorporates new forms or may be incorporated by a form.

But what about Deimel and Baecker, the human designers of these constellations? Was not the assemblage of both forms guided by the concepts and images they had in mind? Should novelty not be regarded as their invention or at least as the embodiment of a new and valuable concept, which is the product of human creativity? Yes and no. Surely, both actors assemble both forms according to their intent to build a specific kind of object. This intent accounts for how the form relates to and/or differentiates from the formats of robotic hands and cybernetic machines. Furthermore, Deimel and Baecker anticipate how parts may work together and assemble them according to their imagined fitting. Stressing these human efforts regards the effective cause (*causa efficiens*) as the reason for a new form. However, such an explanation is somewhat limited, as it may not account for the material agency that makes bits and pieces converge and click in. In both forms, this material agency is efficacious. The RBO Hand creates behavior instead of illustrating a concept, which builds upon the silicone's agency exceeding human governance. Similarly, *Mirage*'s image is a rendering of agencies that precede the assemblage of its parts. Hence, both forms' novelty is an articulation of human and material capacities. For instance, the appropriation of knowledge like technical formats or the tacit knowledge of aesthetics is a human capacity, but, in forms, this knowledge only becomes *effective* when combined with material capacities like the silicone's softness or the muscle wire's contraction and release. Such capacities belong to the material potentials, and less to the cognitive schema in the designer's head (cf. Simondon 2012c; De Boever, Murray, and Roffe 2012). In this sense, I regard articulation as a more appropriate way to explain the novelty of forms than regarding technology as an invention. Articulations stress how heterogeneous elements come together and connect; *novelty of form is then the crossing of thresholds when parts click in and work together as a technical unit.*

3.5 Excursus: Two Techno-Aesthetics

The preceding section describes how a technical form evolves within hybrid constellations. I describe novelty as an articulation of elements that work together and become effective through their circular causality. This articulation of novelty is a pre-condition for sense-making processes, such as applying technological objects in different contexts or re-interpreting the cultural meaning. In the following section, I want to describe another articulation of novelty through technological forms – through their aesthetics. These aesthetics entail moving somewhat away from *causa formalis* toward *causa finalis*, the sense-making of change. Nevertheless, the following account tries to stay in between both causes; it will not dissolve the form into the context of experiencing its affective qualities, yet. The account of aesthetics advanced here remains concerned with the bits and pieces that work together and the modes of their connections.

In an unsent letter to Jacques Derrida, Simondon wrote about "techno-aesthetics," whose primary category is not contemplation, but action and becoming (Simondon 2012c). In his letter, Simondon refers to several examples ranging from simple-and-clever tools, like adaptable cycling wrenches, up to purist designs of viaducts; however, his pivotal example is the architectural approach of Le Corbusier. He referred to Le Corbusier's avoidance of roughcasting as "respect for the material" (Simondon 2012c, 2). Whereas others try to hide the raw construction behind paneling, Le Corbusier integrated redundant material characteristics into a building's design. Simondon called this "phanero-

3.5 Excursus: Two Techno-Aesthetics

technics,"[44] which he also saw in the purist metal designs of the Eiffel Tower and Garabit Viaduct. According to Simondon, these buildings are perfectly functional, successful, and beautiful (Simondon 2012c, 2):

> "It's technical and aesthetic at the same time: aesthetic because it's technical, and technical because it's aesthetic."

He regarded this as the fusion of the technical, which manifests as working and functioning, and the aesthetic, which addresses its affective qualities. In particular, when technologies stray from norms, they enact affective qualities, for example, when a wrench copes with different screws through its seamless adaptability. In contrast to "consumer aesthetics," Simondon saw techno-aesthetics as fundamental moments unfolding in practice, usage, and engineering:

> "It's about a certain contact with matter that is being transformed through work. One experiences something aesthetic when one is doing a soldering or driving in a long screw."

I want to pick up Simondon's remarks and delineate two techno-aesthetics enacted through the RBO Hand and *Mirage*'s form. The first techno-aesthetic is *simplicity*, characterized by the connection of only a few components that work together and thwart more complex technologies. In robotics discourse, this aesthetic is expressed in underactuated designs, which have more degrees of freedom than actuators (Controzzi, Cipriani, and Carrozza 2014, 226-7). The RBO Hand is an extreme example of such designs: it is actuated by only one signal (inflating/deflating) while embodying multiple, innumerable degrees of freedom. The RBO Hand achieves this ratio by exploiting the compliance and deformability of silicone, while equally ensuring enough grasping robustness through its layer design. The aesthetic quality of its form is shown in the performative enactment of the Hand's grasping style in video-stills that illustrate research results instead of theoretical modeling. The Hand does not rely on additional sensors for grasping items, but adapts to the environment through interaction, made possible by the allocation of three silicone fingers and a two-section palm. Furthermore, its assemblage of silicone, twining, and plywood is not hidden but visible; there are no non-functional components assembled in the Hand, whereas other robotic hands implement additional components as, for instance, to make a hand appear anthropomorphic. It is the simplicity of a few components clicking in so as to enact complex grasping movements that gives an affective quality to the Hand. Functionality, not schematic modeling, performs this techno-aesthetic.

---

44 From the Greek word phainó, which means to bring to light, to cause to appear; phanero translates as visible or manifest.

Mirage's form also integrates technical relations characterized by simplicity, such as the muscle wire that connects and actuates two parts. Like the phanero-technics, which Simondon regarded as having an affective quality due to the visibility of technical processes, *Mirage* has an open architecture that lays bare the assemblage of heterogeneous digital and analogue components, its acrylic skeleton, wires, springs, screws, and so on. However, in contrast to the RBO Hand, its architecture is not simple but complex. The *complexity* of manifold components that are functionally related but opaque in terms of their coordination enacts this second techno-aesthetic. *Mirage* performs the aesthetics of complexity through its visual output. Its image renders technical processes that would otherwise remain incomprehensible. *Mirage*'s form integrates multiple components that stem from the constellation of its formation. Technical relations accumulate over the course of its formation in order to find a form that converts unpredictable signals into a fluid movement. The peculiar tension between visible components and contingent output enacts the affective quality of complexity.

This approach to technological forms complements the argumentation above. Not only becoming and causal efficacy characterize the qualities of technological forms, but also their affective qualities. Such techno-aesthetic qualities are not a matter of passive observation and silent contemplation, but entail experience and engagement with technology. This causes them to infiltrate the tinkering with and designing of objects. Techno-aesthetics capture the simple pleasure of making things do something they have not done before – making forms simple or complex, but foremost making them work.

4 Difference: How Categories are Valorized

In the preceding chapter, I delineate novelty as articulated through a technical form. Novelty is the moment when things work together, when parts click in and a material assemblage individuates. However, it is not every scholar's approach to use novelty as a concept for capturing such moments in material practice. The perspective of biographical passages regards novelty as cultural valorization, which is not concerned with material practice but with discourses. Using novelty as a concept to capture valorization is a focus, for instance, of media philosopher Boris Groys. For him, the new is the *valorized* other (Groys 1992, 42 ff.). Whereas otherness and difference do not entail positive acknowledgment as such, in the sense of not enfolding the societal values of an object's significance, novelty includes such acknowledgment of difference as culturally valuable, relevant, and desired. In this sense, his perspective focuses on the difference of differences, as some differences are valued as originality and others remain irrelevant.

In the following chapter, I move toward Groys' concern about novelty as the valorized other and delineate how the RBO Hand and *Mirage* become valorized contributions to robotics and media art. Whereas I have dealt with ideas in Chapter 2 to delineate how the identity of a not-yet-materialized object is articulated, my concern here is how the stabilized technical form signifies its difference in relation and opposition to other existing objects. However, I do not shift entirely to Groys' discursive perspective, which would entail beginning with both objects' public recognition. Rather, I remain concerned with practices and, again, start in the laboratory and studio. This allows tracking of how both objects' difference or otherness *is made* perceivable and of what is selected, enacted, and articulated as novelty.

In contrast, Groys begins in the *archive*. As said, for Groys, novelty is not simply difference, but the valorized other – the other that is found valuable enough to be preserved, investigated, commented on, and criticized in order to prevent its disappearing (Groys 1992, 43). The archive is the crucial marker in this valorization. Whereas, intuitively, archives are not associated with novelty, as they are places where historical documents and records are kept and not produced, Groys has counteracted this intuition and argued that novelty only becomes a positive demand when identity is maintained instead of endangered (Groys 1992, 23). In this sense, archives preserve identity, as they prevent the old from vanishing over time. If no archives existed, societies would need to pass on intact traditions via language and rituals, which would hinder innovation, as the new would come as a threat to tradition, so Groys has said. Archives make identity accessible, as they are societal memories with the ability to store the old that, in return, make a future-oriented society possible. The archive allows for the comparison of cultural products with what came before and the selection of what products mark substantial progress. Hence, novelty can be distinguished from the sheerly different, as it is brought into relation with the old, which has already been captured and valued in the societal memory.

This makes novelty a value that requires a reference in the cultural archive. Referencing is not only possible for authors, but also for their critics, who can equally access the archive and compare the potentially new to what came before (Groys 1992). This referencing allows individual judgment and, likewise, public discussion about the new. In this regard, Groys entails critique as an integral part of valorization. In an essay in which he addressed technical innovations in particular, he noted that critique is inherent to technical innovation and not something external – it is a driver of technology. Groys pledged to regard technological progress not as a temporal linear order, but as a reaction to critique and demands from heterogeneous sites and positions (Groys 1997, 25). This critique relates the potentially new to the repertoire of approved, canonized, and traditional technologies stored in the societal archive.

Groys' archive is a conceptual notion whose pivotal concern is the relation between novelty and the societal or cultural memory. The *abstraction* of the archive-concept is heuristically attractive, as it allows one to use novelty as a concept instead of dropping back to actors' semantics to define what novelty is. Moreover, Groys is very clear about what novelty is: new is what enters the archive (cf. Groys 1992, 44). In his essay on technology, he used the abstractness of his concept to illustrate that valorization through archiving is not limited to the practices of museums and art collections, but is also effective in technological practices such as, for instance, in the sense of patent archives (Groys 1997, 27). Furthermore, and heuristically more important, the abstractness allows for the use of novelty as a concept without necessarily including the actors' judgment of whether the new is morally good or bad. Again, this is clearest in his technology essay, as he specifically addressed that entering the archive enables critical engagement with technology. Entering the archive entails discussing, criticizing, rejecting, or continuing the new in a different form, or that discourse may even valorize opposing life forms. Besides novelty as a conceptual account, I want to stress another positive element in Groys' concept and the relation of old and new: *referencing*. He understands referencing as an activity of authors (artists, engineers, etc.), who reference what is in the archive so as to distinguish between their own product and others *and* institutions (the museum, curators, the patent office, etc.), which reference the archive so as to relate the potentially new to the shared cultural history. In so doing, Groys marks the path into the archive as a multi-sited activity and not simply as recognizing a passive object. This entails referencing as not one clearly defined way into the archive, but as something that can be acted out differently: the logic of the archive dictates what form an inscription into the archive has to take, but not the act of inscribing (Groys 1997, 27).

However, these positive attributes of Groys' archive – abstractness and referencing – are the issues of my critique as well. Groys uses the archive as an abstract concept in the sense of a material and technical cultural memory, but does not consider archives on a lower level of abstraction. He only occasionally gives empirical examples, such as the museum or the patent register, but does not delineate what is specific about archives in, for example, art, science, economy, or politics. This limits the scope of his concept, because it may only account for novelty and not for different forms of novelty. Differentiating what novelty is in different contexts and situations is necessary for tracking how technological objects become objects of science, art, or any other field. Furthermore, this lack of differentiation makes the role of referencing unclear. Groys' account may use referencing to link the old and new, but fails to address what accompanies such activities. This is, for instance, how referencing selects, interprets, and reconfigures the old, as well as how referencing selects what is enacted as the novelty of a potentially new object.

Such selective processes are the concern of science and technology studies. These focus on how laboratory settings actively construct discursive representations of scientific realities. Hence, they trace what is selected by referencing and what this selection enacts as potentially valuable elements of scientific work. This entails stressing the subsequence of forms that translates experimental events into scientific results. One term that captures such translational activities is Latour's "immutable mobile" (Latour 1987). Although Latour did not advance the term toward a profound concept, it is a continuation of arguments concerning articulation and his idea that scientists bridge discourse and form by connecting entities through alliances. As stressed in the introduction of this study, one of Latour's major concerns is not to regard the world and language as a dichotomy in which language simply captures the world as it is. Rather, he proposes regarding the relation of the world and language as a chain of translations from matter to form (Latour 1999, 24ff.). Through articulating world and discourse, the world is always selectively manipulated and aligned step-by-step with the epistemic technologies of scientific discourses. This is, for instance, how Munsell charts (form) classify soil samples (matter), which then enter subsequent procedures as a circulating reference. Hence, world and language, matter and form, are related through alliances of multiple human and non-human actors. Coming from this end of Latour's work, immutable mobiles are a continuation of such chains that connect world and discourse – nevertheless, they entail a shift. This shift is their ability to travel without loss of stability. Latour introduces the term in the sense of a "mobilization of the world" (Latour 1987, 223 ff.). He wrote (Latour 1987, 227):

> "All these charts, tables and trajectories are conveniently at hand and combinable at will, no matter whether they are twenty centuries old or a day old; each of them brings celestial bodies billions of tons heavy and hundreds of thousands of miles away to the size of a point on a piece of paper."

Immutable mobiles, like charts, tables, graphs, maps, etc., translate events into a form that is durable and displaceable. Unlike soil samples, they do not erode, but are stable enough to be transported back and forth, and, hence, become means "to act at a distance," so Latour (Latour 1987, 229). In this sense, immutable mobiles "describe not displacement *without* transformation but displacement *through* transformation" (Latour 2005, 223, author's emphasis). This selective transformation is what Groys' referencing misses. Latour shows how referencing always selects bits of the world in alliance with the epistemic technologies of discourses, and, furthermore, how referencing is a subsequent chain of translating matter into form, and forms into different forms, in order to make events become immutably mobile, and that which may eventually become part of the archive.

However, Latour's translational referencing misses valorization. Latour concentrates on how forms articulate matter and discourse, but he does not include shifts that mark an object's cultural value. This is, for instance, how referencing individuates an object so as to signify its difference from what came before. To conceptualize this, one needs to combine the archive that marks an object's difference from the old with the selective materialization of events that makes references travel. Furthermore, one needs to address how references differ and how they become signifiers of a specific kind. Hence, in the following section, I ask how objects travel into the discourses and how their value is attributed. I attempt to delineate the production of immutable mobiles, how references make objects different from what came before, how discourses reiterate selected elements, and how difference becomes a shared value.

To do so, I use my basic heuristic of articulations that connect figures, technicity, and enactments. Nevertheless, these are very basic assumptions that also signify identity and form as articulations of novelty. Hence, I apply some terminology derived from discussing Groys and Latour, which is significant for the discursive activities of articulating difference as novelty. These concepts are discursive practices, translating, and referencing. *Discursive practices* are shared, standardized, and materialized forms of statement-production, such as press

4 Difference: How Categories are Valorized

releases, scientific articles, and talks (Keller 2007, 64). They are structures that describe the logic of the archive and dictate what form an inscription into the archive has to take (cf. Groys 1997). *Translations* are activities structured by discursive practices. They are linear activities that change something from one medium into another and, in so doing, capture and reify otherwise dissolving events. In this regard, they are basic activities of creating immutable mobiles, entities that represent selected elements of a network (cf. Callon and Latour 1981; Callon 1986). *References* are activities, too, but, in contrast, they create meaning by alluding to similarities and differences between one object and other objects. They can be of different natures: they may differentiate kin concepts, enact shared imaginaries and figures, or evoke meaning through associative similarities of presumably distant objects. Furthermore, referencing is an activity performed by authors as well as critics. Hence, articulating difference as novelty is a matter of bridging the gap between the world and discourse from both sides: from the laboratory and studio side, in the sense of Deimel and Baecker referring to kin objects in their own papers and talks, and from the discourse side, in the sense of critics, colleagues, or institutions perceiving the RBO Hand and *Mirage* and discussing their values in relation to other objects.

A main methodological problem in this section is the ongoing valorization of both objects. By the time I am writing this text, a subsequent version of the Hand already exists and is being increasingly referenced by other roboticists. Similarly, *Mirage*'s journey through exhibitions is surely not finished, yet. Hence, this section is an account of the events that I was able to capture, limited by the economics of my writings. Nevertheless, my comparative ethnography has reached a natural saturation: the RBO Hand has been published, cited, and its descendant awarded; *Mirage* has been exhibited, discussed, and awarded, too. Hence, both authors have channeled their objects into discourses, and these have recognized and valorized their novelty. It is on this tipping point that I focus, instead of giving a comprehensive account of both objects' valorization, whose end cannot be foreseen.

The empirical account begins with sketching selected discursive practices of robotic grasping and media art, related to the valorization of the RBO Hand and *Mirage*. From there, I move back to the laboratory and studio, where I follow how experiments and tinkering practices translate events into immutable mobiles. In opposition to the frequent switching between the RBO Hand and *Mirage* in the previous chapters, this section's structure increasingly separates both objects. This is partly due to the conceptual question, which attempts to delineate different kinds of referencing and archival relations, and partly due to my empirical observations. A pivotal difference in the latter is a scientific experiment conducted by Deimel, for which I did not find a reasonable equivalent in Baecker's practice. As I show, an experiment in robotics is not an open exploration, but is itself a result of aligning form, constellation, and discourse. Standardized procedures structure robotics experiments and inscribe the valorization structure of the field into the laboratory setting. Hence, such differences in empirical observations enforce a change in the structure of my writing and, furthermore, signify that objects become objects of science as well as art. Nevertheless, the accounts of both objects focus on the translation of events and the different kinds of referencing practices. Following the actors' activities, I analyze the recognition of the RBO Hand and *Mirage* in robotics, media art, and public discourses. I close this section by addressing the articulation of difference as novelty.

4.1 Discursive Practices

As mentioned, Groys is very unspecific about archives when it comes to archives outside the art sphere. Furthermore, his focus on museums and libraries seems somewhat outdated, particularly as media art is coordinated through more temporal formats, such as festivals, that stabilize the value of contributions, for example, through awards. Hence, I regard what the relevant archive in robotics and media art is as an empirical question that can be addressed after following the RBO Hand's and *Mirage*'s discursive recognition. Nevertheless, there are standardized ways or channels into the archive that an object has to take in order to be recognized and valorized through a discourse. In discourse analysis, such channels are commonly regarded as *discursive practices*: materialized forms of statement production, such as press releases, scientific articles, or talks (Keller 2007, 64). These prescribe *how* something is communicated and the discursive scenery in which this happens. Hence, it is rather the structure of a discourse that discursive practices describe, not individual actions by actors. In the following, I briefly map such general practices so as to account for the pre-structuring of the channels that the RBO Hand and *Mirage* might take in the course of their recognition and valorization. These practices entail sites such as conferences, exhibitions, festivals, and written publications, as well as graphic representations and iconographies, which reproduce field-specific figures through visual codes. The illustrative examples are chosen according to the RBO Hand's and *Mirage*'s discursive contexts, and most of them play a role in the later sections of this account.

4.1.1 Conferences, Festivals, and Galleries

In robotics, there are plenty of national and international conferences. Some of these are specialized for specific domains, while others are broader or have designated streams for more specific research issues. Within the convolutedness of conferences are a few major international ones that gather contributions from most robotics domains. One of these is the International Conference on Robotics and Automation (ICRA), hosted annually by the IEEE Robotics and Automation Society. These are large-scale events that include industry exhibitions and keynotes by major scientific authors in the field. Contributors present their papers in thematic sessions and often publish these in designated conference proceedings. In 2013, the ICRA accepted 40% of submitted papers.[45] Hence, there is a quality selection for entering such conferences. In general, talks have the typical structure of scientific talks consisting of PowerPoint presentations of about 15-20 Minutes. Despite major conferences such as the ICRA, there are also specialized conferences in small domains such as soft robotics. Soft robotics is a rather young domain that seeks to exploit the potentials of deformable and adaptable materials.

Similarly to the ICRA in robotics, there are also major media art festivals. One of these is the Prix Ars Electronica, held annually as an award ceremony in Linz, Austria, and accompanied by a festival that includes exhibitions of art pieces and conference talks. Since 1979, the festival has awarded outstanding contributions to electronic arts in different domains. Common to media art festivals, like Ars Electronica, Transmediale in Berlin, and the Dutch Electronic Art Festival (DEAF) in Rotterdam, is a conference theme that addresses societal challenges concerning technology and new media, such as "POST CITY – Habitats for the 21st Century," which was the festival theme of Ars Electronica 2015. Thus, it is very common for festivals to appreciate art pieces that comply with a normative statement. Festivals are also sites that capture the variety of media art and its boundaries with electronic music, graphic and new media design, and science. One of such multi-referential festivals is the Fiber Festival in Amsterdam, which hosts media installations, audiovisual performances, and DJ and live sets, as well as workshops and lectures by artists and media researchers. In 2015, the Fiber Festival gathered such various contributions under the theme "The Subterranean – Exploring Networked Tools and Matter," which aimed at investigating the use of digital tools to create, navigate, and excavate "a hidden digital landscape."[46] Besides festivals, there are also regular exhibitions hosted by private galleries, public cultural centers, and museums. The ZKM in Karlsruhe is a museum dedicated to media art that also hosts many activities surrounding the media art discourse. In Berlin, a large media art scene launches bottom up initiatives to exhibit artworks. One of those non-profit initiatives is LEAP Gallery (Lab for Electronic Arts and Performance), which hosts group and solo exhibitions and performances. The temporarily funded gallery hosts a program of new and internationally well-recognized artists despite its lack of institutionalization.

---

45 See the survey under http://www.icra2013.org/ (last accessed May 18, 2015).
46 Website for the Fiber Festival 2015: http://2015.fiberfestival.nl/ (last accessed June 24, 2015).

4.1.2 Journals, Books, and Catalogues

In robotics, articles are a major discursive format for publishing research results. Monographs are less typical for publishing results, but they are more common as state-of-the-art overviews or for meta-discussions (cf. Brooks 2005). Frequently updated robotics handbooks are resources, particularly for teaching (cf. Melchiorri and Kaneko 2008; Balasubramanian and Santos 2014). For publishing the RBO Hand, there are two kinds of relevant articles. Conference proceedings commonly publish contributions and differentiate between short and long papers. Depending on the conference and the hosting institution, proceedings are well recognized in the field and considered a significant research publication. Commonly, conference papers precede the second kind of articles that is relevant to the RBO Hand: peer-reviewed journal articles. As in many other disciplines, these are the major signifiers of what counts as a meaningful scientific contribution. Although robotic grasping is an established and institutionalized domain in robotics research, there are only a few specialized journals. Major robotics journals, like Autonomous Robots and The International Journal of Robotics Research, are publication sites for robotics in general, but also publish special issues on advances in grasping (cf. Ben Amor et al. 2014; Dollar et al. 2014).

In media art, written publications document recent developments or archive media art histories. There are peer-reviewed journals that cover media art as well. There are general art journals such as Art Journal or Journal for Aesthetics & Culture, and journals, like Convergence, that deal with new media in particular. Authors in those journals are not only art historians or media researchers but are also media artists who build artworks as well as write about general issues in the field (cf. Kac 1997; Penny 2000). Furthermore, there are different book formats covering media art. There are, for instance, books mapping the topics that co-evolved with the advent of media art (cf. Wilson 2002; Lischka and Sick 2007), and exhibition catalogues that are retrospectively accounted for as pioneering events, like Cybernetic Serendipity, which covers the 1968 exhibition in London (Reichardt 1968). Designated texts commonly accompany exhibited media art installations and explain the technical concept of an artwork or include a larger narrative to signify the artworks' meaning, like the idea of a hallucinating machine behind *Mirage*.

4.1.3 Graphs, Models, and Technical Drawings

The structure and writing practices in robotics publications entail scientific formalization like mathematical equations or diagrams and graphs. These practices capture formalized models as well as representations of experimental research results. Mathematical models of grasping are formalized so as to capture and predict the behavior of robotic hands and items grasped under the various loading conditions that may arise during grasping (Prattichizzo and Trinkle 2008, 672). They simulate gasping and define ideal grasping types and allocations. Hence, models are benchmarks for evaluating whether or not grasping performance is optimal. They are commonly formalized equations that capture forces and loading conditions, or models as technical drawings that illustrate joints and angles of optimized force control. Nevertheless, models require robotic hardware that matches the undisturbed performance represented in models. Grasping models are theoretical means of representation and are typical for analytical approaches in robotics, while graphs are common ways to represent the empirical results of experiments. Graphs capture grasping performances under various conditions. They capture, for example, tested parameters such as forces or positions, as well as success rates for different items (cf. Dollar and Howe 2010).

In media art publications, graphs, models, and technical drawings are not common representational practices. Art history and media studies, which are typical media art discourses, have a different epistemic and representational culture than robotics or, more generally, the natural sciences, and do not include mathematical formalizations or schematics in their discussions of artworks. Instead, they rely mainly on written texts and images.

4.1.4 Images, Videos, and Visual Codes

What robotics and media art discourses share are manifold images and videos that reproduce the objects discussed. Either these portray objects in a specific way, or discursive practices display them alongside a new object as specific signifiers. Portraits entail *visual codes* that are significant to a particular discourse. Umberto Eco explains that visual codes select specific elements of pictures and relate them to previous experiences (Eco 2002, 202). Hence, they are, on the one hand, signifiers of the referenced object and, on the other hand, visual stimuli that continue in a discourse and relate an object to others in the sense of kinship.

Concerning robotic hands, such visual codes commonly include anthropomorphic elements. This happens, for instance, in images where robotic hands hold mundane items. The affective quality of such images not only addresses lay people, but also scientists. Brock, for instance, recounted how he saw a video of a silicone grabber picking up an egg and, inspired by that thought, he considered that this could also be an approach to tackling the problem of robotic grasping. The visual code of an egg refers to the fragility of its shell; as such, it signifies the particular hand as compliant and sensitive. This code continues in the robotics discourse. In public images, robotic hands with anthropomorphic designs commonly hold fragile items, like an egg or light bulb, by applying a pincer grip. The pincer grip is a force-closure grasp that requires a thumb and an opposing pointer finger. In *Figure 24*, the Shadow Hand[47] performs a pincer grip to hold a light bulb. This kind of portrayal is particularly significant for humanoid robots, since pincer, or precision grips, are thought of as being of a specific human and primate capacity (Butterworth and Itakura 1998). In this sense, the pincer grip transports an evolutionary meaning that signifies the hand as "almost human" (cf. Castañeda and Suchman 2014).

This "almost" or not-quite human is a significant tension for robotics images. On the one hand, hands perform their anthropomorphic abilities, as in the pincer grip or the spreading of their fingers, and, on the other hand, portraits maintain their technological origin, as they show cables, wires, tubes, and other mechanics. However, this is not to be confused with Simondon's phanerotechnical aesthetics, which he regarded as affective qualities of raw and undisguised materials (Simondon 2012c). On the contrary, visual coding in robotics is selective in its reproduction of technical elements. It is only the positively acknowledged parts signifying an object's sophisticated engineering, like joints and tendons, that are displayed in images. The infrastructure needed to drive a robotic hand and that constitutes its technical functioning vanishes from portraits. In robotics discourse, portraits that share these elements are ubiquitous. This is significant for robotics as an engineering science whose concepts strongly relate to their embodiment.

Besides image portrayal, videos of moving robots are standard in robotics discourse. Institutes' or companies' websites and designated YouTube Channels display them to show off a given robot's capacities. The visual codes are similarly selective concerning those of portrait pictures: they usually involve mundane items grasped by robotic hands, and the larger infrastructure vanishes. In robotic grasping, videos commonly reproduce grasping experiments – however, without failures, or only those that can be accounted for as a new research tasks.

Whereas robotics is a scientific field with strong visual codes in its images, media artworks do not necessarily share more visual codes than works from other art fields. It is common sense that art history discusses artworks in terms of their semiotics, which entail specific visual codes and their iconographic meaning. In the case of *Mirage*, I also address aspects concerning the symbolizing relation of its image and the idea behind it. However, I will not move to a semiotic analysis of *Mirage*'s or media art's general visual codes. What I want to mention here are some visual codes concerning the discursive fluctuation of

---

47 Shadow Robot is a robotics company, see: http://www.shadowrobot.com/ (last accessed May 27, 2015).

Figure 24: Shadow Hand holding a light bulb (source, Shadow Robot Company).

media artworks. This includes, for instance, ways of portraying media artworks that share similarities with robotic hands. Similarly to robotic hands, displaying technical elements is common to portraying media artworks, which is shown, for instance, in the online documentation of contributions to the Prix Ars Electronica. The portraits of nominees in the category of Hybrid Art highlight the technical complexity of the artworks. The online slideshows include pictures that capture the full size of installations, as well as close-ups of single elements.[48] Portraits of the winning contribution share these codes: pictures show close-ups of electric sparks and pumping liquids as well as the complete installation spanning the whole exhibition room. Complexity becomes a narrative in these pictures, signified by showing detailed connections and large-scale effects. Furthermore, many contributions are accompanied not only by written text that explains the technical concept and embed an art piece in a larger narrative, but also by technical figures. Such figures are drawings that illustrate the technical workings of an installation. However, it is not merely the content that is significant; it is also the technical character of such schematics and their resemblances to scientific models. They entail reduced information and graphical elements, like vectors for representing technical connections.

Furthermore, moving pictures are common practice in media art discourse. This is, of course, true for graphic animations, where videos are the main medium, as well as for the portrayal purposes discussed above. The jury of the Prix Ars Electronica even states in their submission requirements that their decision primarily considers the video documentation.

The practices summarized above structure how the RBO Hand and *Mirage* have to enter discourses to be recognized and valorized through them. Of course, there are always different ways to do something, and I show that Deimel and Baecker do some things differently. However, significant to the practices sketched above are their roles as markers of a discourse. They have the power to mark the RBO Hand as a scientific object and *Mirage* as an artwork. This is because they reproduce the shared labels of science and art, respective communicative means, and a shared iconography. Nevertheless, it takes effort and action to meet these shared practices. As Latour stresses, the world and discourse have to be bridged step by step (Latour 1999). In the following two sections, I describe how Deimel and Baecker build this bridge by paying special attention to the translation of events and the kind of referencing through which the RBO Hand and *Mirage* relate to particular discourses.

---

48   Website for the Prix Ars Electronica 2015 winners: http://www.aec.at/prix/en/gewinner/#hybridart (last accessed May 27, 2015).

4.2 Translating Events into Other Entities

Discursive practices structure what form an object has to enter to be recognized and inscribed into the archive. Translations are activities that produce such forms, as they reify and fix events triggered by an object's efficacy. They capture and transform things into other media and, hence, freeze temporally alternating processes and states. In this sense, translations enable events as well as plans and intentions to travel beyond the material and spatial contexts of their production. They are able to bridge the material world of the laboratory and studio, including the material and experimental practices that take place in them, with the discourses of a field that attribute an object's value.

4.2.1 The RBO Hand: Translating Events into Charts and Pictures

My account begins with the translation of events that take place in the laboratory. The first of such translations is an experiment conducted by Deimel. I want to stress at this point that I selected for this account empirical incidents that are significant to the articulation of novelty. The actual beginning of the building of a chain between an object and a certain discourse could hypothetically lie elsewhere. In the case of the RBO Hand, the chain certainly precedes the setting up of an experiment. Nevertheless, the experiment is a significant point of passage for producing immutable and mobile entities that meet the discursive practices in robotics. After delineating two kinds of failure within that experiment, I turn to how pictures enact the RBO Hand. In focus is what is hidden and concealed through the production of clusters, graphs, or pictures.

Graphs from an Experiment

> "The grasping movement is simply performed in every position. Then we make a cluster from them. Of course, such work is time consuming." (Deimel, RBO Hand)

In Sections 3.2 and 3.3, I address the hybrid constellation in which the RBO Hand's form emerges, as well as the Hand's distinctive kind of grasping enacted within that constellation. The constellation described there is similar to the experimental set-up addressed in the following, in terms of its composition of technical elements. Whereas I focus on the Hand's technical functionality in Chapter 3, in the following, I want to analyze how the constellation becomes an experimental setting that produces events that make the Hand's grasping a recognizable contribution to robotics.

To describe the hybrid constellation in Section 3.2.1, I address the work needed to align the Hand's specificities with the Meka robotic arm and its steering program. There, I report on a scene wherein Deimel prepares an experiment for which he programs arm movements that drive the mounted Hand into a suitable position for grasping. The scene that I turn to now continues with the actual experiment a few days later, for which I revisit Deimel in the laboratory (*Figure 25*). In front of the Meka arm and torso with the mounted RBO Hand stands a table with a tube wrapped in plotting paper. Deimel explains that the experiment's aim is to test different degrees of air pressure and how these change the Hand's behavior according to different grasping angles. Hence, he has preprogrammed test runs that automatically run several different arm movements in order to vary the angle at which the Hand approaches the tube. These runs structure the experimental setting according to a sequential order of grasping events. Every grasping movement is documented according to its success. The collected data is then clustered (cf. quote above) to visualize what air pressure and angle constellation is most efficient. The transcribed scene shows Deimel sitting beside the table in front of the Meka ensemble. A test run is currently running. The scene (*Figure 25*) entails two kinds of failure that appear frequently during the whole experiment.

The first kind of failure is *legitimate failure* and is shown in Frames I and II. Legitimate failure requires an experiment setup that works reliably. When the setup works as planned, the Meka arm approaches the bottle at an angle that allows the Hand to potentially grasp the tube successfully. Deimel intends this kind of movement for the test runs. During the whole sequence, Deimel focuses his gaze on the arm movements. Although the arm works as planned, in Frames I and II, its movement is not smooth but rather slightly stifled. The first frame shows how Deimel positions his body in front of the bottle. From the beginning of the sequence, he holds his own arm in a position that allows him to intervene. The Meka arm moves toward the tube, and the Hand bends its fingers. As the grasping is not strong enough, the tube slips out of the closing Hand. Deimel intervenes and prevents the tube from falling off the table. Despite the Hand's failure to grasp the tube, the arm continues its movement: it lifts the Hand and then drives back to its default position. Deimel briefly waits for the arm to fulfill the programmed movement and then places the tube back onto its designated spot. Significant to legitimate failure is that it is not a surprising event. Deimel anticipates that some grasps will be unsuccessful and performs this through his bodily attention to the sequential order of the experiment. He is ready to intervene, which cause him to prevent the bottle from falling off the table and to replace it in time to continue the experiment without interruption. This shows how Deimel is bodily involved in the working hybrid constellation. Legitimate failure

4.2 Translating Events into Other Entities 175

	Time	Still of the video	Transcript
I.	25:43	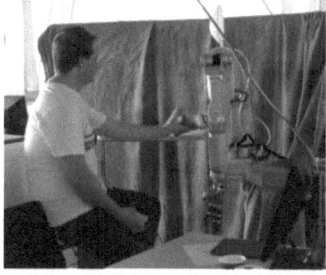	Deimel sits beside the ensemble of RBO Hand, Meka arm, and torso. The robotic arm drives the RBO Hand toward the bottle, the Hand grasps, and the bottle falls. To prevent the bottle from falling off the table Deimel grabs it. The arm continues the programmed movement and lifts the Hand without the bottle.
II.	26:00	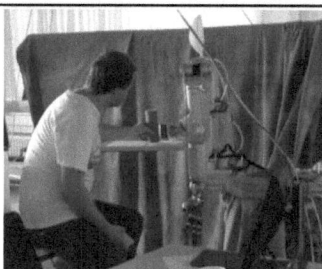	The arm drives back to its default position. Deimel replaces the bottle onto its designated spot on the table.
III.	26:10	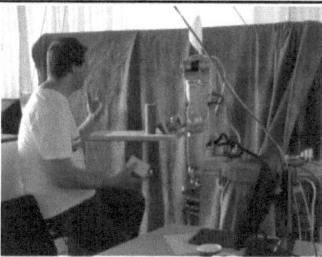	The arm continues with the subsequent movement. It drives the Hand into the table. Deimel immediately pushes the emergency button. He lifts his arm and sighs.
IV.	26:20	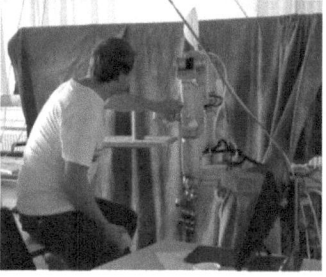	Deimel grabs the wrist with his left hand and pushes the emergency button again with his right. Thereupon the arm continues the movement and lifts the Hand without bottle; Deimel keeps holding the wrist. He pushes the red emergency button. The arm stops its movement and Deimel stands up and takes a seat in front of the computer screen.

Figure 25: Two kinds of failure in an experiment with the RBO Hand (own video).

maintains the working constellation and does not break the sequential order of the experiment. Furthermore, the setting translates it into a graph so as to document it as an unsuccessful grasping event. This failure is legitimate, because the infrastructure works, and, hence, the varying alignment of air pressure and grasping angle are accountable for the failure of the grasp.

The second kind is *illegitimate failure*, which follows in the next sequence in Frames III and IV. During illegitimate failure, the experimental setup does not work as planned, due to incorrect alignment of the Hand, Meka arm, and steering software. The sequence shows how the arm sinks unexpectedly and drives the Hand into the table. From that position, there is no successful grasp possible, as the movement's angle is too steep and the arm collides with the disturbing environment. Deimel immediately pushes the emergency button to stop the run and prevent the arm from incurring damage. Pushing the emergency button shows that illegitimate failure might not be desired, but it is anticipated. Deimel is constantly holding the button in his hand, ready to use. In contrast to legitimate failure, which Deimel bodily anticipates, the constellation implies technical entities that allow the interruption of running processes in anticipation of illegitimate failure. Deimel indicates that he did not expect illegitimate failure, despite his precautions, and lifts his arm and sighs. In Frame IV, he grabs the arm's wrist with his left hand and pushes the emergency button again. He keeps on holding the wrist as the arm continues its movement and lifts the Hand without a bottle. By holding the wrist, Deimel feels the directions of the arm movement and the applied torque. After guiding the arm back to its default position, he pushes the emergency button again, which stops all processes. Deimel gets up from his position during the running experiment and sits down in front of his computer, from where he can rework the steering software. The illegitimate failure illustrates how connections of multiple agents (software, the robotic arm, the Hand, the table, Deimel) enhance the contingent behavior of the constellation. Hence, having installed a running experimental setup does not necessarily mean that it works without technical interruptions caused by components other than the object under investigation.

The whole sequence shows that the experiment is not an open exploration, but a scripted procedure performed according to scientific standards. Its aim is not to create behavior as such, but translating differential patterns that occur through the controlled changes in the experimental system into a stable and immutable entity. Controlling those patterns means enacting changes in a sequential order that can be captured and fixed as data (cf. Rheinberger 1992). This does not mean the hand has to grasp successfully in every setting. On the contrary, unsuccessful grasps are a constitutive part of experiments, as they enact significant differences occurring through controlled changes. The immediate succession of legitimate and illegitimate failure shows how fragile experiments are and that control is more a matter of creating documentable results than stabilizing installed relations.

Comparing the observed experimental procedure with the published results (or "clusters") exemplifies how these mobilize the laboratory world as a highly purified version in research papers. The results of the experiment are translated into graphs that distinguish between different grasping allocations and the probability of their success (Deimel and Brock 2013, 2043). Success areas are marked as light squares, which turn darker towards the unsuccessful grasping allocations, which are black (*Figure 26*). Black squares are legitimate failures. Illegitimate failures are not shown. The published charts are immutable mobiles in Latour's sense, which bring "celestial bodies billions of tons heavy and hundreds of thousands of miles away to the size of a point on a piece of paper" (Latour 1987, 227).

But what do they hide? It is not the experimental system as such that vanishes, as the graphs are accompanied by a text that describes the experimental setting, including the Meka arm and torso. Rather, *the enactment* of the experimental setup vanishes in publications. This is the trial-and-error work needed to align the RBO Hand with the Meka ensemble and the steering software, as well as Deimel's bodily involvement during the whole experiment. His involvement, like preventing the tube from falling off the table, maintains the sequential order of the experiment, but vanishes through the translation of grasping events into graphs. In contrast to the enactment of order in the laboratory, which is broken, interrupted, and re-adjusted, the sequential order of test runs is re-enacted through the charts as a continuous process that renders grasping as the purified product of location and air pressure.

Grasping Pictures

I have already indicated that the RBO Hand's functioning somewhat exceeds modeling. Deimel describes this as "non-linearity" of the actuators' movements caused by the silicone's material characteristics. The limited modeling of the Hand's functioning results in a large share of pictures in Deimel's and Brock's research paper on the Hand. The paper, which also includes the graphs above, includes several pictures that give an account of the Hand's movements. These pictures are taken in the same laboratory setting as the experiment. Some of these pictures are rows of video stills that show the fingers' bending under different air pressure settings (Deimel and Brock 2013, 2041, Fig. 3). Others are shots from a fixed perspective, showing the Hand grasping different items (Deimel and Brock 2013, 2044, Fig. 6). Furthermore, there is a row of video stills that concludes the paper and shows in detail how the Hand proceeds for a surface-constrained grasp, which is regarded as the kind of grasp the Hand is particularly good at (cf. *Figure 22*) (Deimel and Brock 2013, 2045, Fig. 9).

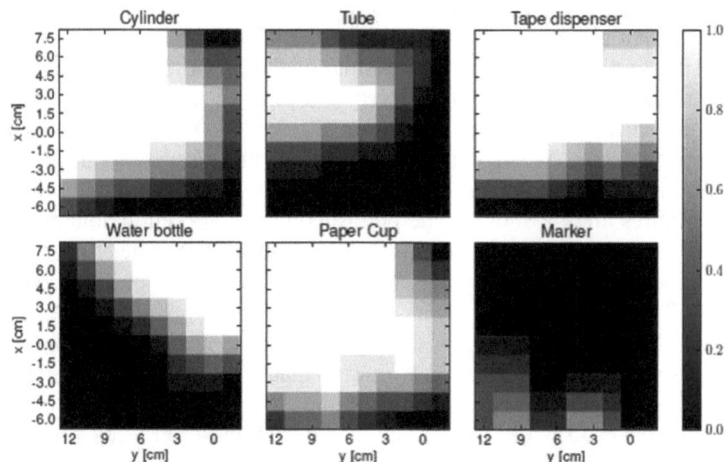

Figure 26: Graph translated from grasping experiment (source, Deimel and Brock 2013, 2043, Figure 7: Sliding grasps success probability under object placement variation).

These kinds of pictures enact the Hand's distinct movement, as they translate selected events of experimental conduct. They differ from the charts in so far as they focus on the material characteristics that are explicitly *not* translatable into graphical representations. Rather, the pictures are a way to enact the distinctive working of the Hand as they dynamically portray what the Hand does. The reader does not need to understand a representation of the Hand, like deciphering a technical model or graph, but sees the Hand's way of grasping.

Nevertheless, these pictures are selective. Similarly to the charts, they hide something in return. This is partly a logical consequence of zooming in. However, more significant is the concealing of the experimental constellation, which consists not only of the partly shown Meka arm, but also of the wires, the desktop computers, the steering software, the emergency button, and Deimel, who conducts the ensemble. In this sense, the archival technology of pictures selects what the archive can recognize as the RBO Hand's distinctive grasping.

4.2.2 Mirage: Translating events into Drawings, Pictures, and Video

As mentioned in the beginning of this section, empirical differences in how Deimel and Baecker channel the RBO Hand and *Mirage* into their discourses partly forms this section's structure. A central difference is the scientific experiment, which translates the RBO Hand's grasping into graphs, and for which I could not identify an analogous procedure in *Mirage*'s enactment. This is a significant difference in terms of immutable mobiles produced. Nevertheless, my account of translating *Mirage* into other entities that match the discursive practices of media art, also begins at the site of the object's material assemblage, the studio. The translation of events in the studio is not as directed towards translating results, but rather is a matter of bridging *Mirage*'s form with the media art discourse. After describing how Baecker translated a not-yet-realized version of *Mirage* to acquire funding, I continue with the documentation of *Mirage*'s final form. Nevertheless, there are two very different reasons for documenting *Mirage*' s behavior: firstly, to align *Mirage*'s technical components, and then for a submission to a call of the Prix Ars Electronica.

Drawings

I have already addressed that Baecker anticipated exhibiting *Mirage* from the beginning of its design. In Section 3.2.2, for instance, I describe how future scenarios materialize in the studio as test structures or by imitating a dark exhibition room. However, that concerns how Baecker brought the media art field into the studio, whereas, in the following, I want to address how he brings *Mirage* into the media art field.. The threshold to be crossed in the design of *Mirage*, which caused Baecker to engage with the institution of the field, came after stabilizing some technical features of *Mirage*. For instance, by that time, Baecker tested ways how to enact a moving laser light image and how to create a skeleton of acrylic glass that could hold muscle wires. After having tested these elements, Baecker was somewhat sure how to realize his idea of a hallucinating machine.

One of the first engagements between Baecker and media art institutions regarding his new installation *Mirage* was an application for funding. The application addressed LEAP Gallery, which acquires public funding to support the realization of media artworks that they later exhibit in their space. Baecker and LEAP Gallery had been engaged before, so both parties knew about the other's work. However, to apply for funding, knowing each other is not enough, as the ambition of building an artwork requires a basic, fixed agreement about the funded piece. LEAP Gallery's means are partly the reason for this necessity, as they are not private but public assets and require legitimization. Furthermore, the future artwork's form needs to be comprehensible, and something needs to indicate that its realization is advancing and is not only an empty promise. Media artworks' material specifics are crucial information for galleries, as different media entail very different exhibition requirements – and *Mirage* has large spatial requirements.

Baecker's application for LEAP Gallery's funding is a pdf-file that consists of photos, technical drawings, and a short text. Its first page is a photo of a curved red laser projection in a darkened environment described as an image of the prototype. Additional photos show the mirror foil and parts of the acrylic skeleton. The technical drawings are already close to *Mirage*'s final design (*Figure 27*). All such drawings are computer drawn images with brief descriptions of the components. They are three-dimensional so as to display *Mirage*'s spatial dimension and, hence, spatial requirements for a future exhibition. Two drawings show the metal frame, the laser diode, and the projection, all assembled according to their potential spatial order from varying perspectives. The anticipated exhibition room is not arbitrary here: *Mirage*'s spatial order in the drawing shows that it requires considerable space between the empty wall, on which the image is projected, and the metal frame, in order to display the image. Two other drawings are detailed images of the frame's multilevel architecture. The short text, which is included in the application, is a brief description of *Mirage* as a projection apparatus and about how the technical components relate to the moving image.

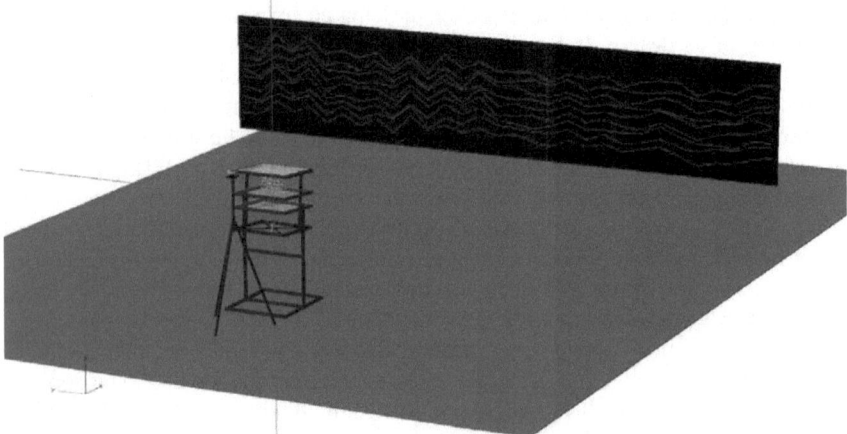

Figure 27: Technical drawing of *Mirage*'s design approximately six months prior to the exhibition (source, Ralf Baecker).

The overall application translates inquiries into an application for funding that had taken place in the studio. At the time he was applying, photos illustrated the progress of the project, and the technical drawings fixed Baecker's reasonable objectives. Nevertheless, the drawings are considerably different from the translated graphs of the RBO Hand. Instead of representing the efficacy of an object, the drawings point into the future. They anticipate the future exhibition context and, as such, the discursive practices of media art. The technical drawings translate Baecker's explorative tinkering in the studio as well as his design concept into a potential contribution to LEAP Gallery's exhibition and to the media art field. Eventually, the application was successful; LEAP Gallery supported *Mirage*'s production and hosted its first exhibition in April 2014, approximately half a year after application.

Pictures

"I have seen so many states; I can hardly tell which was best." (Baecker, *Mirage*)

Baecker moved *Mirage* to its first exhibition space in LEAP Gallery almost one week prior to the opening. He was aware that the spatial conditions would influence the image rather drastically and, hence, wanted to adjust *Mirage* early in the exhibition mode. After he set up *Mirage*, he started fine-tuning the image. I have already addressed the means of doing so: changing the height of the laser diode, focusing the laser, skipping the inner frame, and re-positioning the double-sided tape that connects muscle wire and foil of acrylic glass. One of Baecker's major concerns was to get rid of the dints caused by the connection between muscle wire and acrylic glass. Doing so was a matter of calibrating the position of the tape, the muscle wire's tension, and the three-dimensionality of the image changes by skipping the frame.

When I visited him during the setup, I noticed a camera in front of the image. Baecker told me he had brought his camera to document how the image appeared so far and how it changed based on his calibrations. He said that he changed the setting so many times and had seen it in so many states that it was hard for him to tell whether a new calibration was better or worse than a previous one. In this regard, the camera was a means to monitor his own activities so as to find an appropriate system mode for the exhibition.

However, on the day of the opening, Baecker also took pictures. By then, the room was tidy, and no packaging materials or tools were lying around – the calibrating was done. In addition, the darkening of the room was optimal: the image projection was visible on a wall at the completely darkened end of the room, and the installation frame was at the other end, where a little light was shed on its structure. Baecker took pictures in this final setting from different perspectives. He not only portrayed the image during calibration, but the whole exhibition setup. He took pictures from a wide range of perspectives, showing the complete frame and the image, as well as close-ups of the muscle wires, the mirror of acrylic glass, and the Arduino boards. He told me that it would take a lot of effort to setup *Mirage* like this again. Thus, he wanted to exploit this optimized situation and took pictures of *Mirage* for his own documentation. Later, these pictures would appear on his website and, since he uses them for press releases, published articles have reproduced them as well (cf. *Figure 3*).

Video

Before Baecker actually concretized *Mirage*'s form, he had already built the first step to bridge his future installation and the media art world, by applying for funding. However, the bridging continued after having finished the piece. To answer the annual open call for submissions to the Prix Ars Electronica, Baecker selected some of the pictures he had taken at LEAP Gallery and, additionally, created a video that shows *Mirage*'s moving image and hardware elements. The submission guidelines for the category Hybrid Art explicitly require a maximum of one video for each installation.

Mirage's video begins with the moving laser image.[49] The video is set to a subtle feedback sound. For one and a half minutes, the video shows how the laser projection appears as a three-dimensional moving image: it only shows the red laser light on a black surface. In the next sequence, the video begins to show *Mirage*'s hardware parts alongside an explanatory text. First, the fluxgate magnetometer is in focus, and the text explains that it is registering the Earth's magnetic field. Then, the video shows the small display of the Raspberry Pi with quickly changing numbers. The text goes on and states that, simultaneously, an algorithm is generating versions of the previously perceived signal. This blends into a video-screenshot of the converting algorithm, which was not actually part of how *Mirage* was exhibited, but added to the video. After approximately two minutes, the video takes a wider perspective and shows the complete metal frame and the acrylic skeleton. Again, text accompanies the pictures, indicating that "the hallucinated landscape data" actuates 48 muscle wires that deform the shape of a mirror foil. The video and text continue with the laser diode. The text explains that a diode sends a laser beam onto the mirror's surface so as to throw "a landscape-like projection on the wall." After these explanations, the video continues with approximately 45 seconds of the slowly changing hardware parts of *Mirage*. This is followed by one minute of the moving laser image, like at the beginning of the video. At 4:30 minutes, the video ends with a reference to LEAP Gallery as supporters of *Mirage*'s production.

---

49   The pictures and videos of all the winners and honorary mentions are publicly accessible. *Mirage*'s entry can be found at: http://prix2015.aec.at/prixwinner/15446/ (last accessed June 23, 2017).

The video, including visuals and text, is selective. In focus are *Mirage*'s several hardware connections and the conversion of signals. This selection is particularly significant, because the video adds screenshots of the converting signals, which is usually not part of the exhibited installation. What is hidden from the video is the alignment of the installation according to the exhibition surroundings and, simply, the alignment to what Baecker regards as the best state. In this sense, the video enacts the technicity of *Mirage* and selectively translates its contingent and complex movements into an entity to be potentially appropriated by the media art discourse.

4.2.3 Stabilizing, Concealing, and Travelling of Relations

The preceding empirical descriptions entail very different translations of the RBO Hand and *Mirage*, but they share distinct similarities, too. The translations are different regarding the range of forms they bring up, which spans graphs and videos, as well as in terms of their situated production. For instance, the experiment with the RBO Hand requires a standardized scripting of sequences, which enactments of *Mirage* do not show. Despite these empirical differences, translations share certain features that are the first bricks in building a bridge between the laboratory and studio world and the robotics and media art discourse.

All translations *stabilize temporally dissolving events* or object states. This is most evident in the experiment sequence that illustrates how fragile hybrid constellations are and that controlling the production of differential patterns is rather a matter of producing results than stabilizing a technical order. Deimel maintains the sequential order of the experiment bodily so as to fix legitimate grasping events into data clusters and graphs. These stabilize the messy enactment of the RBO Hand as a purified product of position and air pressure. In a very different setting, pictures of *Mirage* stabilize the frequently changing states at the exhibition site. Baecker begins fixing states of the moving image when he realizes the sensitive reaction of the system state to changing environmental conditions. The tidy and optimized setting of LEAP Gallery is a temporally dissolving event that he wants to exploit for producing pictures as public representations of *Mirage*. In this regard, translations produce entities that are stable representatives of otherwise dissolving events.

Furthermore, *translations select and conceal relations*. The graphs that result from the experiments conceal Deimel's bodily involvement in conducting the experiment and select those events that research papers can reproduce as scientific results. It is not the constellation as such that papers conceal, but especially the profanity of producing results that includes Deimel's body as well as software, wires, emergency button, etc. – things that are taken for granted but still constitute the experimental order. The selectiveness of translated events is most evident in comparing the video sequence of the experiment and the published representation of its results, which translates robotic grasping into a two-dimensional product. The selective concealing of situated enactments is not limited to graphs but continues in the translation of objects into pictures. The documentation of *Mirage* conceals its messy environment and reproduces the laser image in its temporally optimized aesthetics. This selects temporal states as representations of *Mirage*'s technical workings. In this sense, translations select and conceal the material relations of an object's enactment so as to produce a tidied representation of their technical efficacy constituted by several other profane entities.

Finally, translations make events *travel* beyond the contexts of their production. This is what Latour has understood as the folding of events into a new form, which allows displacement *through* transformation (Latour 1987). The graphs published in Deimel's first paper illustrate this. The graph is an entity that allows the representation of events in the laboratory as research results reproduced in a paper-format publication. Other researchers can access the resulting pdf-file or printout through their pdf-readers or libraries no matter where they are located. Baecker's documentation at the exhibition site included this aspect, too, as the pictures appeared on his website and, from there, have been accessible as mobile representations of *Mirage*. However, Baecker's translations bring up an aspect not explicitly addressed by Latour: mobile entities may also include anticipations and, hence, point to future enactments. The technical drawing of *Mirage* translates the tinkering practices in his studio into an entity that already represents its future exhibition context before Baecker actually realizes the installation. The technical drawing is a mobile representation of a future enactment and, in this regard, produces an anticipatory entity that bridges the studio world and the discursive practices of media art.

4.3 Referencing by the Authors

In contrast to the concealing of relations through translations, the following section deals with *adding* relations and the enactment of *references*. Like translations, references are activities to signify an object. In contrast to translations, they create meaning by alluding to similarities and differences between the object and other objects, concepts, imaginaries, or figures. Nevertheless, they do so in varying ways and through different substrates: through writings, images, videos, as well as through material entities such as technical components (Stubbe 2015, 121ff.). In the following, I concentrate on Deimel's and Baecker's referencing practices and how both allude to the similarities and differences of other objects, concepts, figures, and imaginaries. I begin with Deimel's referencing practices, followed by associative references that relate *Mirage* to diverse realms.

4.3.1 Referencing the RBO Hand

Referencing practices concerning the RBO Hand entail portraits, written publications, and talks, as well as a video tutorial in which Deimel explains the easy manufacturing of the PneuFlex Actuators.

Portraits of the RBO Hand

Besides the grasping pictures, there are plenty of other public pictures of the RBO Hand. In contrast to the dynamic rows of video stills, several pictures are portraits that show the Hand in a fixed position. The paper mentioned above have published some of these, while others have been published on the institute's website and advertising posters. As mentioned, a high share of visualization is typical for robotics, in particular if research implies engineering hardware. There are roughly three kinds of portraits of the RBO Hand.

The first kind is made up of still lives.[50] I have already made use of one still life for *Figure 1*. That picture shows the fully assembled RBO Hand in front of a clean background. The Hand is mounted on the Meka arm, whose wrist is visible in the picture. The picture is published on the RBO laboratory's website as well as in the research paper that I have already cited (Deimel and Brock 2013, 2039, Fig. 1). It is the first figure in this paper. Besides the portrait that I have used, the figure includes a second picture of the RBO Hand with spread fingers, showing its inner surface. The still life does not entail the typical anthropomorphic visual codes of robotics portraits. However, like in other common portraits of robots, the picture shows those technical parts considered positive trademarks. This is the RBO Hand's inner technical structure, which is open and visible, and its connection to a larger robotic system as indicated by the wrist of the Meka arm.

The second kind of picture is also made up of still lives, but ones that show only single parts of the Hand. In the paper, these show, in particular, the PneuFlex Actuators and their exact composition. They are shown so as to illustrate the finger's manufacturing process and lay open how the passive and active layers are glued together, as well as how the air tube is channeled into the passive layer and the helix twining wound around it.

All of these pictures highlight the Hand's assembled form, its detailed finger composition, and that the Hand fits into a robotics infrastructure. In so doing, they reiterate visual codes that are common in robotics. One code is the displaying of technical elements, which selectively opens technical designs. Portraits of

---

50 A still life is a work of art portraying mostly inanimate subject matter.

the RBO Hand are zoomed in when it comes to the inner composition of the PneuFlex Actuators.

Writing about the Hand's "Novelty"

"A Compliant Hand Based on a Novel Pneumatic Actuator." (Title of Deimel and Brock 2013)

I have already cited the first of Deimel's and Brock's papers on the RBO Hand several times (Deimel and Brock 2013). I use it as a source to give detailed accounts of the Hand's technical features as well as to account for what the authors regard as its novel output. The paper is part of the conference proceedings of the ICRA 2013. Generally, it follows the standard structure of scientific articles: introduction, method, results, and discussion. However, the paper includes some sections that do not quite fit that structure. Its first level headings are Introduction, Related Work, Actuator Design, Manufacturing Process, The RBO Hand, Experimental Results, and Conclusions. The structure describes the actuators and the RBO Hand separately. It gives a lot of space to the Hand's fingers, the PneuFlex Actuators, as the paper goes into a description of their design as well as manufacturing process. The section headlined "The RBO Hand" is an account of the actuator's implementation so as to assemble them into a functioning form.

Furthermore, the paper entails semantics of novelty. Considering the paper's structure and title, it is obvious that the authors consider the PneuFlex Actuators as *the* focal point of the approach's novelty. Their novelty is distinctively described in contrast to an already established actuator, the PneuNet Actuator (Ilievski et al. 2011). Deimel and Brock have written that the PneuNet approach is easy to manufacture, but is limited by the small ratios in achieved elasticity. In order to overcome this limitation, their "novel PneuFlex Actuator embeds polymer fibers to reinforce the rubber substrate" (Deimel and Brock 2013, 2040). It does so for two reasons: firstly, in the passive layer fibers enable bending without significant elongation. Secondly, the reinforcement helix along the entire actuator creates anisotropic elasticity in the active layer (cf. Section 3.3.2). The section enacts the PneuFlex Actuators' novelty as an improvement through a new manufacturing method and design. The approach builds upon existing attempts to use soft materials for grasping, but creates a distinctively different efficacy through its improved material characteristics. Through experiments that are described in the remainder of this paper, this improvement is signified as a contribution to robotic grasping, because the PneuFlex Actuators can be implemented as a robotic hand, whereas the PneuNet Actuators were only manufactured as experimental forms. Concluding the paper, Deimel and Brock explicated the claim that their novel actuators are a contribution to robotics research (Deimel and Brock 2013, 2045):

> "We believe that this novel way of constructing hands could lead to simple and competent end-effectors for mobile manipulation."

However, the paper enacts novelty not only as an undoubtedly good improvement. Instead, novelty also refers to the preliminary and experimental state of their research. Introducing the paper, Brock and Deimel wrote (Deimel and Brock 2013, 2039):

> "We emphasize that the RBO Hand described here is a first prototype based on a new design objective, a novel actuator, and a novel manufacturing process."

They continue and state that further research could extend the concept and use it for other grasping techniques, but this has not been done yet. By emphasizing the prototype state of the RBO Hand, Deimel and Brock encourage the reader to assess the value of their contribution according to its potential and less regarding grasping abilities already implemented. They use novelty as a term that puts their work into perspective, so as to prevent assessments according to grasping standards alone. In a conversation, Deimel confirmed this and told me that, to some extent, their approach's novelty and originality makes up for a lack of grasping versatility. In the paper as well as in the clarifying conversation, Deimel used novelty as a semantic form that qualifies a relation. This relation is between a materialized unit, its abilities, and its potentials on the one hand, and the field to which it aims to make a contribution and that field's established criteria on the other hand. Novelty semantically qualifies this relation as valuable – not only according to *what is*, but also according to *what could be*.

Referencing Kin Research

Of course, scientific articles need to stress the state-of-the-art of a research domain in order to relate their work to relevant literature. This is not only a matter of validating an argument, but, furthermore, a way to address differences.

In their article, Deimel and Brock stated that the use of shape-matching and compliance is not novel to robotic grasping as such. In a designated section, they related their Hand to a number of compliant grippers (Deimel and Brock 2013, 2039-40). They began with the SDM Hand, which also implements an underactuated design but, in contrast to the RBO Hand, is based on rigid fingers (cf. Dollar and Howe 2010). Then they reported on a Pneumatic Anthropomorphic

Hand, which uses compliant actuators, too, but, according to Deimel and Brock, its "rigid finger links and hinge joints" reduce "its ability to match the shape of an object during grasping." The next reference is a gripper similar to the Starfish Grabber. The authors mentioned its similar actuators, but stressed the 5-10 times lower air pressure that results in weaker grasps than those of their RBO Hand. The following reference is also a compliance-based gripper, but one that does not allow the grasping of freestanding items. Deimel's and Brock's last reference is the Awiwi Hand. They referred to the Awiwi Hand as "probably the most capable anthropomorphic robot hand built to this day." It achieves compliance through a tendon-driven, antagonistic actuation system. However, they stressed, it is mechanically complex, expensive, and requires sophisticated control. In contrast, they "pursue an opposing philosophy" (Deimel and Brock 2013, 2040).

Referencing kin research signifies *compliance* as a relevant characteristic of robotic hands. In this sense, the approaches referenced show that compliance leads to improved grasping performance. However, each reference also entails a critique that enacts what the RBO Hand does differently and/or better. The first two references stress the RBO Hand's enhanced compliance, as its actuators do not include any rigid elements. These actuators are, in contrast to both of the following references, strong and allow robust grasping of not only lying but also freestanding items. Through the RBO Hand's distinctive and simple design, this is possible by using cheap material that does not require sophisticated control like the superior Awiwi Hand does. Hence, referencing kin research enacts the RBO Hand as more compliant, robust, versatile, and simple in opposition to its kin robotic hands.

Human and Silicone Fingers at the ICRA

"This is what we want to mimic." (Deimel, RBO Hand)

In the following, I want to report on a presentation of the RBO Hand at the ICRA 2013 in Karlsruhe. The presentation was part of session entitled "Hand Design." All contributions were paper presentations and followed a standard mode of 15-minute PowerPoint-supported talks, of which approximately three to five minutes addressed questions from the audience. Whereas the conference hosted an exhibition that demonstrated several different robots, the single sessions did not usually entail hardware demonstrations. Nobody brought a robot along to a talk in order to demonstrate a research approach. I want to highlight two significant aspects of Deimel's talk:

First, Deimel started his talk with a *video of a human hand grasping* a sponge from a table (*Figure 28*). He referred to this as the role model of their approach. The hand in the video grasped the sponge by spreading the fingers and capturing the sponge between the tip of the thumb and the middle finger. The hand repeated this grasping action several times, while Deimel commented that, at first glance, this is a very ordinary thing for humans to do. However, he went on, when looking at the video in slow motion, details become visible that are crucial to human grasping. The video switched to slow motion, then paused, and two circles popped up highlighting the fingertips. By visually focusing on the fingertips, one could see that these not only were capturing the sponge, but were also touching the surface of the table and, hence, were slightly bent backwards as they slid over it. Deimel explained that human grasping exploited the finger's compliance and used the table as resistance to pick up the sponge, which was the reason for success. Before continuing with the technical part of the talk, he said, "This is what we want to mimic."

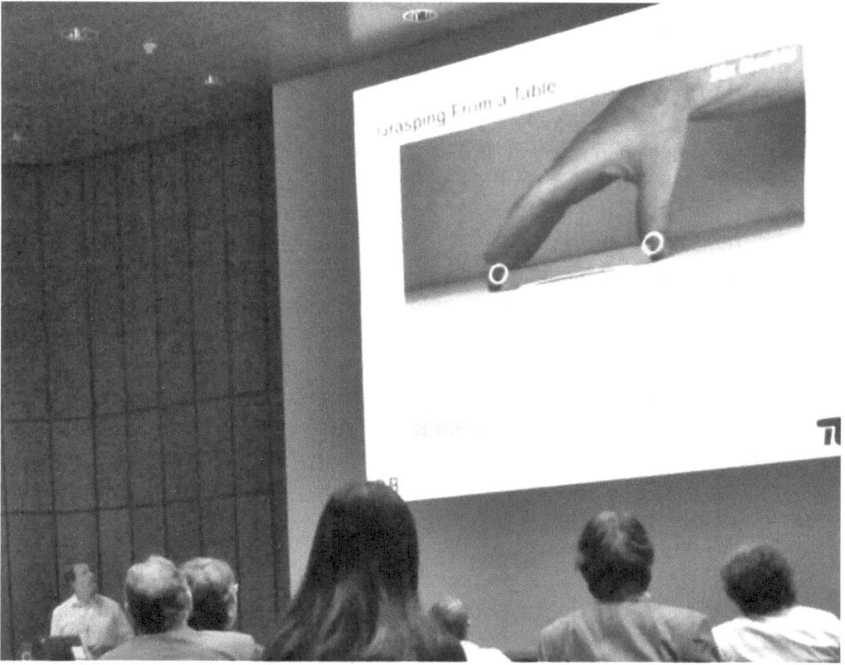

Figure 28: Video-screening of human hand grasping a sponge at the ICRA (own picture).

In this sequence, Deimel gives a similar figurative account of human grasping as the text on the institutes website does, which I report on in Section 2.2.1. That text gives an account of the human capacity to grasp a cup without consciously activating all senses. The text figures specific capacities of human grasping as relevant criteria for robotic grasping. The video sequence above renders the same capacity; this time, not as a tropic story, but by enacting a videotaped human hand. The video is very selective: it does not show the arm that moves the hand into position, nor the human eyes that are probably looking at the sponge. In this sense, the video materializes a selected event of human grasping and enacts this as a general criterion for robotic grasping. By stating that this capacity is to be "mimicked" in the RBO Hand, Deimel reiterates the human figure as an ideal. Reiterating the anthropomorphic ideal connects the RBO Hand's approach to other papers presented in the session and, furthermore, to the general humanoid imaginary.

The second aspect of Deimel's talk that I want to highlight occurred after he started reporting on the state of the art of compliant grasping and explained that their approach attempts to use silicone as a new material for grasping. Unlike other presentations that went on without hardware presentations, Deimel brought along single fingers from the RBO Hand to *demonstrate their material capacities*. The silicone of the RBO Hand is clearly a distinct feature of its design. I have already mentioned that the Hand's fingers have their own label as PneuFlex Actuators, which indicates that, according to the actors involved, the fingers significantly embody the RBO Hand's concept – they are somewhat if a figurehead for the whole project. During his talk, Deimel picked up one PneuFlex Actuator, including a designated air pump, and inflated the silicone finger with air, whereupon the actuator bent. He held the actuator and pump in one hand and pointed upwards to his slides with his other hand, while looking at the audience. Then he moved from his position behind the speaker's podium and passed around three PneuFlex Actuators with hand pumps. The attendees were able to touch and inflate the single fingers and try out the ways in which they bent. During the presentation, I was also sitting in the audience. When the actuators and the air pump were passed along to me, I realized that their basic functioning could be demonstrated rather easily. Everybody was able to make the finger deform and simulate grasping.

This interactive presentation materially referenced the concepts mentioned in his talk. In this sense, Deimel enacted the material explorations in the laboratory that make up an integral part of their scientific work but that are usually absent at a conference talk. However, Deimel also created a new event. He not only talked about concepts and visually coded criteria, but also created a haptic experience of the silicone's characteristics. Its compliance and deformability provided an evocative experience, as the fingers did not embody the pragmatics of a complete grasp, but the silicone's potential.[51] Attendees could feel the actuator's distinct behavior without necessarily understanding it in the sense of a scientific model. The actuators are bendable and compliant, and they interactively respond to their environment by deformation, as opposed to robotic hands made out of solid material, which need to be precisely programmed in order to grasp – many of the latter were presented in other papers of the session. The silicone's specific agency signifies the hand as distinctively different from the established categories in its field. Like what writing "novelty" does semantically, the single fingers not only materialize the present state of *what is*, but become agents within the RBO Hand's construction as novelty, as their material evokes thoughts of *what could be*.

Video Tutorial about how to Build PneuFlex Actuators

> "We will teach you a technique to easily create highly compliant actuators and literally soft robotics." (Deimel, introduction of video tutorial)

The selective focusing on the RBO Hand's actuators as novelty continues on the institute's website and YouTube channel. Deimel created a video tutorial on how to produce a PneuFlex Actuator. The video begins with the quote above, followed by: "they are cheap, customizable, and robust." Deimel starts by physically enacting the actuators' characteristics by bending, twisting, squeezing, and inflating a single finger. Then, he goes through the manufacturing process step by step. He introduces the necessary components and tools, then shows how to produce the casts, mix the ingredients for the different layers, cast and glue them together, and twist the helix twining around them so as to make the actuators bend in a directed manner. Deimel's explanations are easy to understand and include only a few scientific remarks. In total, the video takes approximately 13 minutes.

I mention above that Deimel and Brock wrote in their paper that they "pursue an opposing philosophy" to the technically complex Awiwi Hand (Deimel and Brock 2013, 2040). In the video tutorial, Deimel performs this philosophy by physically going through the manufacturing process. He reveals the actuator's simple but effective design and encourages the audience to build their own exemplar. In so doing, he not only speaks and writes about the Hand's compliant concept but also *performs* the concept.

---

51 For the evocative potential of material objects, see Sherry Turkle's *Things We Think With* (Turkle 2007).

4.3.2 Referencing *Mirage*

Similarly to Deimel, Baecker's practices also entail textual, visual, and material references. The writings, talks, and pictures surrounding *Mirage* draw connections to diverse realms and not to media art alone. Thereby, references draw on shared imaginaries that allude to peculiar connections and similarities.

Materialities

During a conversation in his studio, Baecker told me about his choice of materials. I have already used part of that conversation to explain the pragmatic effects of choosing a laser, acrylic glass, and digital components as materials to work with. Baecker told me that his usual choice of materials starts with such pragmatic considerations as having the means to work with it. However, there is something more to it. Baecker recounts how he began working on his previous installation, *Rechnender Raum*. For that particular piece, he worked with wooden staffs, strings, plumb weights, and small digitally programmed electric motors. The installation forms a large structure, approximately two meters in both height and diameter. The wooden staffs serve as a frame that holds together a complex, three-dimensional mesh of strings. Through the plumb weights and pre-programmed motors, the mesh moves in such a way that the spectator cannot exactly tell what is causing the deformation or where the strings are going to move next. Baecker explained that his intent to use wood as the main material was purely pragmatic in the beginning. It was cheap and easy to work with. Later on, he explained, he became aware of the fact that the assemblage of wooden parts, strings, and plumb weights is a reference to mechanical technologies of the 17th and 18th centuries. This reference, which was not intended by him during the initial conceptualization of the work, was re-produced through an invitation to install the piece in another exhibition that was celebrating the foundations of modern science.

Baecker is aware that the materials he uses for his installations carry such semiotics. This is also the case for *Mirage*. Similarly to the wood used for *Rechnender Raum*, the combination of acrylic glass, metal, and visible electronics evoke references. Baecker has said that, for him, acrylic glass brings along the aesthetics of a scientific experiment. This is an important analogy for his art. Such material references indicate the figures of his artworks that draw on forgotten scientific technologies.

Writings

Writing is an activity that accompanies Baecker throughout his artistic work. Through texts, he communicates the specificities of his installations and the ideas behind them. In his published texts, he has used references to other technologies in an associative way – linking imaginaries. This is, for example, how he has referred to the Helmholtz machine, which introduces a sleep-wake metaphor to artificial neural networks (Hinton et al. 1995). Baecker does not refer to it as a concept that he aimed to embody through *Mirage*'s aesthetics, but as a figure that can be used and re-arranged. This figure, not the concept, is a signifier of his dreaming and hallucinating machine.

Baecker told me that he comes across those stories through his research. Often, the technology fascinates him, and then he starts to dig into a particular direction. Sometimes he tries to rebuild mechanics or basic technologies with the materials he has at hand in his studio. Other times, he writes things down. These are often only fragments that he has in mind, but he knows that, at some point, he will need a storyline to signify his objects. However, he told me, those stories become coherent only retrospectively. He collects stories and materialities in parallel over the course of his creative process and, by advancing a technical piece, he sees how story and materiality fit together.

Although there is not strategic foresight involved, Baecker's own writings become closely related to his artworks. The text that he wrote for *Mirage* is published on his own website and in the first exhibition flyer and was circulated as a press text that became the basis for discussing the piece in media art/technology blogs and journals.

Talks

> "When one understands the causes, all vanished images can easily be found again in the brain through the impression of the cause. This is the true art of memory…" (Rene Descartes)

In May 2015, approximately one year after the first exhibition of *Mirage*, Baecker gave a lecture at the Fiber Festival in Amsterdam. The festival announced the lecture as an artist talk about Baecker's general artistic approach and about his recent work *Mirage*. According to the announcement on the festival's website, the first part of the talk was entitled "Lines, Ropes and Computational Spaces," in which Baecker took the audience on a "walk-through of his previous works that show the paradoxical relation between algorithmic thinking and physical matter."[52] The second part of his talk was more specific and presented Baecker's research behind *Mirage*. The website indicated that Baecker would "explain his interest in geological data and landscape, and machine learning and dreaming."

Baecker's slides of the second part of the talk begin by citing Descartes. It is a quote from Descartes' *Cogitationes Privatae* (see above). Although this quote does not entail the word "dreaming," it still continues the trope of a hallucinating machine, as it is concerned with images and the virtuality of the brain. In his notes for the talk, Baecker commented:

> "For me a Computer is not just a calculating machine, but a tool for amplifying imagination, making it possible to weave structures of pure abstract symbols and see them rendered as concrete things."

This comment is programmatic for Baecker's work, as it addresses his pivotal interest in the entanglement of virtual and physical realms and, furthermore, the rendering of abstract or algorithmic processes through physical things. He continues in the presentation by referring to the Helmholtz Machine by Hinton and explains its functioning as a kind of dreaming algorithm. After these introductory references, Baecker gets into more detail concerning *Mirage*'s technical components. He begins with the Earth's magnetic field, which the fluxgate magnetometer captures and transmits as data. He clarifies that he tried various implementations before working with this data source. After several slides of how scientific technologies render magnetic fields in general, as, for instance, through mapping solar winds or in geophysical surveys, Baecker moves on to the physicality of Fata Morganas. One of the slides included a historical picture of an apparatus that aimed at imitating Fata Morganas. He showed that they are caused by diffracting light reflections. This notion of reflections took Baecker's slides to the use of deformable mirrors in astronomy, where technologies use adaptive optics to reduce optical aberrations caused by atmospheric turbulence. After these scientific references, Baecker continued with the technical setup of *Mirage*, indicating that he had built his own deformable mirror system to modulate beams of light to generate an image.

This talk is significant for Baecker's way of presenting his work. For previous talks, which addressed different installations of his, he used such multiple scientific references as well. In this talk, he began with a philosophical quote and a similar reference to artificial neural networks. Both references continued the tropes whose enactment I delineate in Section 2.2.1. Still, they were more concrete in this talk, as Baecker combined philosophy and AI and, hence, illustrated

---

52 Website for the Fiber Festival's announcement: http://2015.fiberfestival.nl/festival/programme/ralf-baecker/ (last accessed June 24, 2015).

the peculiar relation of physicality and images by connecting quotes from two different realms. In the following slides, his references turn from AI to physics, geology, and historical optical science and toward astronomy. He draws connections between these according to their similar concerns for rendering virtual processes – that is, the similarity that connects the references to his artwork *Mirage*. Nevertheless, he does not claim to adapt all these concepts for his art, but states that these are references that evoke one to think about how technologies render hidden processes. This is not a conceptual transfer from science to art, but the drawing of connections between *imaginaries of science* in order to evoke similarities *and* deviation in contrast to the common understanding of appropriating scientific concepts. In this regard, Baecker's *references do not signify the validity of an argument as in science, but articulate* Mirage*'s difference through associative similarities.*

Website, Twitter, and Tumblr

The associative referencing of Baecker's talks and writings continues in his online activities. Baecker runs his own website[53], which he mainly uses for public representation of his work. The first screen of the website juxtaposes all of his publicly exhibited artworks. There are no nametags on this first page, only photos of the installations. Clicking on each photo links to a page designated for each installation with explanatory texts and additional pictures. *Mirage*'s homepage displays the pictures from LEAP Gallery as well as a text, which is a refined and edited version of the text about the hallucinating machine discussed in Section 2.2.1. The text ends with an external link to an online article about *Mirage*, which I address below, and a credit to LEAP Gallery for their support. The website also contains selected references to Baecker's public recognition. There are pages designated for his current activities, biography, exhibitions, and entries in catalogues and books, as well as recognition by relevant online press.

Furthermore, the website contains a link to Baecker's Twitter account[54]. Whereas the website is concerned with Baecker's own work and designed similarly to a CV, Baecker's Twitter account contains more pictures of different scientific technologies other than his own. Certainly, Baecker retweets when other online authors mention his work, and there are also pictures of the exhibition of *Mirage* on his profile, but his own tweets mainly include references to his research process and less to his finished work. Some of these pictures show technical tinkering sessions between him and his befriended colleagues or stu-

---

53 Baecker's website: http://www.rlfbckr.org/ (last accessesed June 25, 2015).
54 Baecker's Twitter account: https://twitter.com/rlfbckr (last accessed June 25, 2015).

dents. More frequently, Baecker tweets pictures similar to those in his talk. His account is full of technical diagrams, portrayals of ancient scientific apparatuses, computer diagrams, historical models from biology, and even a geological profile of the Alps, which renders sediment layers as a folded and deformed horizontal line (*Figure 29*). This profile is exemplary for the associative referencing that continues on Twitter. Conceptually, *Mirage* has nothing to do with the Alps, and *Mirage* is not mentioned in the tweet. However, by juxtaposing pictures of *Mirage* and the profile in the sequential order of his Twitter profile both are connected by the similarity of their shapes. The similar image of folded lines appears in peculiar kinship to distant forms whose only connection is the referencing of Baecker's Twitter account.

Figure 29: Profile of the Alps from Baecker's Twitter account (tweeted on March 16, 2015).

Such pictorial associative referencing continues on Baecker's Tumblr.[55] His Tumblr's appearance focuses even more on historic scientific images than the Twitter account. That is, whereas the Twitter account also includes retweets and links to current events, Baecker's Tumblr has no cross-links and, hence, only displays his picture collection. Still, the pictorial references share the visual codes of his presentations and tweets, such as ancient laboratory settings, technical drawings, or primitive renderings of computational processes. Like on Twitter, pictures of *Mirage* appear in the row of such images, juxtaposing his own work and peculiar forgotten technologies. This juxtaposition signifies Baecker's work as concerned with the aesthetics of technologies. It does so without addressing this concern as a textual explanation, but by associative similarities evoked through the juxtaposition of pictures.

---

55 Baecker's Tumblr: http://rlfbckr.tumblr.com/ (last accessed June 25, 2015).

4.3.3 Conceptual, Figurative, and Associative References

References add relations between an object and discourses. In this sense, they create meaning by alluding to similarities and differences between the object and other objects, concepts, imaginaries, or figures. They do so through varying practices, substrates, and media. Concerning the RBO Hand and *Mirage*, these range from pictures, writings, talks, and demonstrations to the semiotics of certain materialities. Despite the variety of media, three kinds of referencing practices dominate the discursive enactment of the RBO Hand and *Mirage*: conceptual, figurative, and associative referencing.

Conceptual referencing dominates Deimel's writings on the RBO Hand's distinct style of robotic grasping. Writing "novelty" signifies the RBO Hand, and in particular, its PneuFlex Actuators, based on their potential to contribute to shared problems of robotic grasping. Deimel's and Brock's paper entails conceptual references to existing robotic hands, which highlight similarities to and differences from archived robotic technologies regarding their technical features and implementations. "Compliance" is the focal point of those references, which signify the term as a concept that has a history within robotics research as well as potentials that have not yet been realized by other robotic hands. At the ICRA conference, compliance was additionally enacted through the sample fingers that made the silicone's different grasping style possible to experience (Stubbe 2015, 121ff.). In this regard, conceptual referencing alludes to similarities and differences between two entities' essential features, which determine their application or technical implementation.

Figurative referencing accompanies the RBO Hand as well. The videotaped human hand rendered a version of human grasping that reiterated the human figure as an ideal for robotics and, simultaneously contested previous figurations, as it focused on a presumably neglected aspect of human grasping, that is, the fingertip's sliding over the table's surface. In this regard, figurative referencing builds upon shared figures of a discourse, while simultaneously reiterating the figure, creating stability and contending with previous figurations. In this sense, figurative referencing is normative, as it alludes to specific characteristics of a shared figure and enacts these as more appropriate, neglected, and better than previous enactments of the figure.

Associative referencing dominates Baecker's writings, talks, and online activities while also accompanying his choice of materials. He implements certain materialities that evoke similarities between his artworks and ancient scientific technologies. These associations do not necessarily advance a new project, but come together bit by bit through the assembling of technical forms and his research about scientific apparatuses. His talks and writings connect these loose narratives and tell a coherent storyline that refers to the philosophical, scientific, esoteric, and aesthetic concepts that signify *Mirage*. Nevertheless, Baecker does not refer to concepts in the sense of essential features that determine applications or technical implementations; rather, he indicates similarities of diverse scientific realms that build upon shared imaginaries of science and the aesthetic similarities between technologies. In contrast to the figurative referencing in Deimel's practice, associative referencing does not reiterate the given figures of the discourse it contributes to and, furthermore, does not render a presumably more appropriate or better version of an imaginary or figure. Rather, associative referencing draws connections between different realms and discourses and alludes to their similarities. In this sense, associative referencing draws connections between distant imaginaries in order evoke their similarities, *and*, in this regard, it signifies how an object builds upon and simultaneously deviates from the common understanding of these imaginaries.

Despite the difference of these referencing practices, they share a similarity: they converge at a few focal points. These have already been articulated in other sections of this study, but here they become meaningful as discursive statements of both objects' differences. In terms of the RBO Hand, this is *compliance*, which is the key element that continues throughout all of the Hand's references. It is signified through conceptual and figurative references as a meaningful difference from other robotic hands. Furthermore, the PneuFlex Actuators embody compliance as a concept, and Deimel and Brock enact them as the trademark element of the Hand by describing them in detail in the paper and tutorial as well as physically demonstrating their characteristics at the conference. In terms of *Mirage*, the focal point is the *hallucinating machine*. Baecker's associative references signify the figure of the hallucinating machine across diverse realms that share a concern with the visual rendering of hidden processes. This concern holds together the otherwise loose references that he enacts in his writings, talks, and online activities concerning *Mirage*. Both focal points, that of compliance and the hallucinating machine, share that they embody both objects' differences in opposition to other objects, concepts, figures, or imaginaries.

4.4 Referencing by the Archive

In the following, I change perspective: the focus is on what I have so far called the discourse. In this regard, the following account turns toward Groys' archive and the valorization of objects as novelty. I have already addressed that Groys' archive-concept is limited so far as it remains abstract and does not entail archives, which are more specific for the contemporary science and art discourse. To overcome this impediment, I have opened the empirical section of this chapter by addressing specific discursive practices that are not the activities of single actors but are shared approaches to discourse production and materializing statements. This focus on practices allows the alignment of Groys' institutional and Latour's active perspective and, in so doing, regard archival valorization not only as an institutional mark, but also as being produced through references by multiple actors. From there, one might find practices that are specific for valorizing both objects as novel contributions to robotics and media art, without dropping back to the archive alone as an institutionalized form of novelty.

This conceptual trick also takes the empirical problem that I mention earlier into account. This is that both objects' recognition is ongoing, and their valorization is not yet exhausted. Addressing archival valorization as practice instead of institutional mark alone allows one to regard this empirical problem not as a lack of data, but as significant for both objects' valorization. Valorization, from this perspective, is not the ascribing of a fixed value to an object that is stable throughout its existence in the archive, but is a matter of actively selecting what is relevant about an object and re-enacting this within specific temporal discursive practices and, furthermore, *doing* something with it. I begin this section by sketching the recognition of the RBO Hand through the scientific robotics discourse as well through public media channels, followed by *Mirage*'s recognition in media art discourse and other technology-related publications.

4.4.1 The RBO Hand in Citations, Awards, and Online Newspapers

Up to the time of this study, three discursive practices have recognized, discussed, and valuated the RBO Hand. These are citations, awards, and online newspapers.

Citations

Deimel and Brock published the results of the RBO Hand for the first time in a paper from 2013, as already mentioned several times. The following analysis is based on a sample of publications citing that paper. It is not a comprehensive citation analysis, as, presumably, there are additional citations to come. Hence, I go through four exemplary papers from 2014-5, of which three are refereed conference proceedings and one is a peer-reviewed journal article. All of them contribute to robotic grasping research.

The authors of the first example aim at promoting their own robotic hand through their conference paper. That hand is claimed to be open-source, affordable, modular, light-weight, and underactuated (Zisimatos et al. 2014). Introducing the paper, the authors stress that complex and anthropomorphic designs increase the cost of robotic hands, which leads to an unbalanced cost-efficiency ratio. Hence, they pledge for a low-cost open source design. Designs with such premises have lately been developed based on elastomer material; they report on and refer to the RBO Hand as one example of such designs (Zisimatos et al. 2014, 1).

The second conference paper tackles an issue conceptually related to the design of the RBO Hand. The authors proposed an approach to planning for hands that are simple, in the sense of a low number of actuated degrees of freedom, and soft as being continuously deformable throughout their interaction with items. Logically, in order to stress the importance of developing the planning for such hands, the authors need to show that soft manipulation is a recent research concern and has proven its potentials. Hence, they referred to the RBO Hand, which is "designed to be much simpler, and much more robust with respect to the whole interaction process" (Bonilla et al. 2014, 581).

The conceptual relation to the RBO Hand is similar in the third example. The authors are concerned with exploiting environment constraints for grasping (Salvietti et al. 2015). I show in Section 3.3.7 that surface-constraint grasps are the RBO Hand's specialty. Instead of creating behavior, which is the design approach of the RBO Hand, the authors seek to model such grasping techniques. Hence, they proposed "a mathematical representation of robotic grasping in which a compliant hand exploits the environment surface to reach the object in a reliable and robust way" (Salvietti et al. 2015, 8). Introducing their approach to modeling they referred to the RBO Hand, which "uses a novel pneumatic actuator design in its fingers" that makes the hand "inherently compliant" and "highly adaptable." They stressed that underactuated and passively compliant hands, such as the RBO Hand, guarantee robust grasping performance under sensing and actuation uncertainty. However, according to the authors, it remains unclear how to design hand/wrist stiffness to enhance robotic grasping and manipulation capabilities; this is why they want to model grasping (Salvietti et al. 2015, 1).

The fourth citation example is taken from a peer-reviewed article that was published in a special issue of the journal Autonomous Robots on autonomous grasping and manipulation (Ben Amor et al. 2014). In the article, the author's presented "a low-cost, soft cable-driven gripper, featuring no stiff sections, which is able to adapt to a wide range of objects due to its entirely soft structure," while equally ensuring a "stable and safe grasp" (Giannaccini et al. 2014, 93). Hence, their research objective is very close that of to the RBO Hand and concerned with hardware development. As this example is a full journal article, the authors had more space to discuss different approaches, like the RBO Hand, from which they seek to differentiate their own gripper. Firstly, the authors referred to the RBO Hand and the concept of "shape match" (Deimel and Brock 2013, 2039) so as to emphasize the importance of compliance for robotic grasping, which permits grippers to conform its surfaces without explicit control and sensing. However, according to the authors, "Once the grasp is established, an end-effector must be [...] stiffer than it is in its completely compliant initial stage" (Giannaccini et al. 2014, 93). The RBO Hand, as in Deimel's and Brock's 2013 publication, lacks such a dynamic stiff mode. Secondly, the authors are not convinced by the RBO Hand's surface; while they positively acknowledge it as ensuring a good shape match, they stress the difficulty of fitting sensors on it (Giannaccini et al. 2014, 94).

All four articles address their own distinctive research objectives. In so doing, they pick up different characteristics of the RBO Hand and its design approach: its low production costs (Zisimatos et al. 2014), its distinctive grasping style (Salvietti et al. 2015), its sensor-free design (Bonilla et al. 2014; Giannaccini et al. 2014), its general simplicity (Zisimatos et al. 2014; Bonilla et al. 2014), and, most repetitively, its compliance, which allows shape matching (Salvietti et al. 2015; Bonilla et al. 2014; Giannaccini et al. 2014). The authors refer to these characteristics so as to signify their own work and relate it to recent research concerns in robotic grasping. Commonly, references entail positive acknowledgements, while subsequently stressing impediments. The research presented in the journal article by Giannaccini et al. is closest to the RBO Hand and, hence, most specific about its impediments. Similarly to how Deimel and Brock referred to the limits of other robotic hands in their paper, Giannaccini et al. have referred to the RBO Hand so as to enact the difference of their gripper and its progress in combining softness and stiffness. Whereas Deimel and Brock and Giannaccini share their concern with hardware development, Bonilla et al. and Salvietti et al. have referred to the RBO Hand so as to signify their research in a different realm: grasp planning and grasp modeling. Curiously, these are exactly the two realms that the RBO Hand contradicts, as its design opposes planned grasping and theoretical modeling. Nevertheless, the authors refer to the Hand to signify their own research objectives. All these citations reference the RBO Hand in terms of their own research concerns; they select relevant characteristics of the RBO Hand and re-arrange them according to a new context.

Awards

"Originality." (RSS evaluation criteria 2014 and Polanyi 1962)

Large-scale robotics conferences commonly include "best paper" awards in several categories. These often correspond with research domains such as manipulation, vision, and motion. Additionally, there are "best student paper" awards or "best conference paper" awards. The ICRA 2013, where Deimel presented their paper, also awarded papers. The natural category for their paper was the "manipulation" award, which aimed at highlighting "innovative efforts in the planning and execution of manipulation tasks, which take place in dynamic environments."[56] However, Deimel's and Brock's paper did not win an award at that conference.

Nevertheless, a subsequent paper of theirs did win a "best student paper" award at the 2014 "Robotics: Science and Systems Conference" (RSS) in Berkeley, CA, USA. In that paper, Deimel and Brock reported on an advanced version of the RBO Hand, which they call RBO Hand 2. The advancements include a change in morphology toward a more anthropomorphic five-finger design, as well as improved capabilities in the Hand's dexterity (Deimel and Brock 2014). Although the Hand looks considerably different, the authors still place the PneuFlex Actuators in focus, basic functionality of which is still close to the 2013 paper (cf. Deimel and Brock 2014, 2-3). At the RSS in Berkeley, Deimel's and Brock's conference paper received the award for best student paper. The general evaluation criteria for all categories, evaluated in a blind review process, include:

1. Technical strength: Is the paper technically sound?
2. Evaluation of results: Are the claims well supported (by experimental evaluation or proofs)?
3. Significance and relevance: Is the community likely to use the results?
4. References to prior work
5. Clarity: Is the paper well organized and clearly written?
6. Originality: Does this work contain new problems or approaches? Does it combine existing methods in novel ways?

One of the conference organizers explained to me in an e-mail that they had asked reviewers to give scores according to these criteria, which were then summarized and compared. She wrote that Deimel's and Brock's paper had the highest score in the student paper category and was additionally evaluated positively in a discussion by five non-public senior researchers.

Although the previous paper by Deimel and Brock from 2013 did not win an award, the list of evaluation criteria is significant, as it illustrates the valuation structure in robotics. The listed criteria are more specific versions of Michael Polanyi's three general criteria for scientific contributions: plausibility, scientific value, and originality (Polanyi 1962, 57-58). The RSS criteria are more specific – firstly, as they explicitly address professional standards of robotics' epistemology, such as experimental methods and proof, and, secondly, as they stress clarity in writing, which relates to discursive practices as short papers that require clear-cut language.

---

56   Website for the ICRA 2013 describing the award categories: http://www.icra2013.org/?page_id=153 (last accessed July 2, 2015).

Furthermore, the criteria mirror Deimel's translating and referencing practices. The first two criteria are inscribed in the standardized experimental procedure, which translates the RBO Hand's grasping into a scientific representation. Criterion 3 is met by citations that show how the robotics community refers to the published results. Respectively, Criteria 4 and 5 structure Deimel's referencing practices, which enact distinctions of the RBO Hand in opposition to other robotic hands. Finally, Criterion 6, "originality," mirrors the RBO Hand's basic constitution as a research project, which started explicitly as a design approach to create new behavior instead of theory and as combining soft robotics and grasping. Additionally, Criterion 6 mirrors "novelty" as a semantic category applied by Deimel and Brock in their paper from 2013 to indicate the potentials of the RBO Hand.

In this respect, the RSS Award valorizes the combination of the silicone's compliance and robust grasping and its implementation in the RBO Hand through originality as evaluation criteria. Polanyi explained in this regard that originality might overlap with the systematic importance of a discovery, but the surprise of an original discovery causes admiration for its daring and ingenuity. "It pertains to the act of producing the discovery" (Polanyi 1962, 58). Whereas plausibility and scientific value tend to enforce conformity, valorizing originality encourages dissent and difference.

Online Newspaper Articles

"Deutsche Wissenschaftler entwickeln sanfte Roboterhand." (DPA)

At the end of June 2013, the German Press Agency (*DPA*) released an article about the RBO Hand entitled "German Scientists Develop Gentle Robotic Hand" (quoted above). Several German online newspapers, such as *Die Welt*, *Wirtschaftswoche*, *Handelsblatt*, and *Berliner Morgenpost*, published the article.[57] *Spiegel Online* released a slightly modified version of the *DPA* release.[58] Shortly prior to the arti-

---

57 *Die Welt*: http://www.welt.de/print/welt_kompakt/print_wissen/article117526268/Deutsche-entwickeln-sanfte-Roboterhand.html
Handelsblatt: http://www.handelsblatt.com/technik/forschung-innovation/mit-druckluft-deutsche-wissenschaftler-entwickeln-sanfte-roboterhand/8420806.html
Wirtschaftswoche: http://www.wiwo.de/technologie/forschung/innovation-deutsche-wissenschaftler-entwickeln-sanfte-roboterhand/8425906.html
Berliner Morgenpost: http://www.morgenpost.de/newsticker/dpa_nt/regioline_nt/berlinbrandenburg_nt/article117509886/Berliner-Wissenschaftler-entwickeln-sanfte-Roboterhand.html
All articles were released between June 27 and 30, 2013 and last accessed on July 2, 2015. The same press release can be found in several further online newspapers.
58 *Spiegel Online*: http://www.spiegel.de/wissenschaft/technik/weiche-roboterhand-luftdruck-laesst-silikonfinger-greifen-a-908224.html (released June 27, 2013, last accessed July 2, 2015).

cles' release was the 2013 RSS conference in Berlin, which was probably the occasion for the press release, although this was only mentioned in the *Spiegel Online* article. The articles were accompanied by a picture showing the RBO Hand mounted on the Meka arm and holding a paper coffee cup.

The text highlights the RBO Hand as a recent and ongoing research activity at TU Berlin. It explains that the actuators are made from silicone and actuated by air pressure, which makes them soft and adaptable to the environment. The *Spiegel Online* article's headline describes the Hand as a "novel robotic hand, which is more versatile, cheap, and simple to manufacture." (*Figure 30*)[59] It refers to the Hand's "fingers, which are made from silicone, whereas the role of the muscles is taken over by air chambers." The main part of the text begins by addressing the Hand's simplicity, while allowing precise and complex grasping movements. It stresses that the Hand is easy to manufacture and can be reproduced by others due to its low costs; this is encouraged by the TU-researchers, who posted instructions for manufacturing the Hand online. The text quotes Deimel, who stresses that the Hand does not need sensors, and, furthermore, does not damage an item's surface, because of its softness. Shortly after that quote, the text is interrupted by a subheading in quotation marks: "'Helping Humans in Everyday Life.'" The *DPA* article, published, for example, in *Wirtschaftswoche*, includes the same quote and continues by addressing the "vision that one day robots might help humans in their mundane activities." Again, the article quotes Deimel, who explains that robots could, for instance, carry things from A to B, search for lost keys, or tidy up rooms. However, the article continues the quote by saying that robotic hands will not reach human dexterity anytime soon.

---

59 An empirical side note on "novelty": Below the *Spiegel Online* article is a section for user comments. That section includes three comments that all express doubt in the Hand's novelty. The first comment stresses that such "rubber fingers" had already been introduced by a company at an exhibition in Hannover 25 years ago; sadly, I cannot find that company. The second comment stresses a product that I was able to find online. The comment refers to the company Festo, which has a "pneumatic muscle" in their product range. That product is not a robotic hand, but a machinery part for industrial manufacturing facilities. The third comment refers to "Prof. Tanaka from Yokohama University," who, apparently built similar grasping tools. I could not find that professor's work, maybe due to spelling mistakes or the like. What is interesting about those comments is that they all criticize the label "novelty" and do so by stating that all this has been done before. In so doing, they select specific elements of the online article, whether the Hand's rubber material, its pneumatic actuation, or simply the word "muscle," which is phrasing by the journalist. In these comments, novelty is a value-laden category that entails a connotation of absolute novelty, in contrast to an understanding of novelty as recombination. However, a new combination does not seem to justify the label "novelty" according to the commentators.

4.4 Referencing by the Archive

Figure 30: Screenshot of the *Spiegel Online* article on the RBO Hand (taken on July 2, 2015).

The article selects specific elements of the RBO Hand. Again, the Hand's material design is the focus, which is simple but allows complex grasping movements. Furthermore, the article addresses its easy manufacturing process as a difference from common robotic hands, which are technically complex and costly. The article acknowledges the community efforts the researchers make by putting a manufacturing manual online. All these elements narratively connect specific characteristics of the Hand to larger narratives of creativity and design, which are positively connoted in technological discourses (cf. Suchman 2011b).

Furthermore, the articles reiterate the figures of humanoid robotics through the visual codes entailed in the picture showing the RBO Hand holding a cup.

The concurrence of the scientific robotic hand, robotics infrastructure, and a mundane item, which is immediately identifiable as part of everyday human life, is typical for robotics images and echoed in the picture here. It renders the RBO Hand as an object of science, signified by its open technologies, and it equally projects the Hand into a future everyday setting. The text continues about the figure of the picture as it refers to scenarios of robots as "helpers," especially in tasks that do not require being totally human, as they are profane or unpleasant. Nevertheless, the text does not construct scenarios so as to regard the Hand as humanlike. On the contrary, in a quote, Deimel emphasizes that science is, by far, not even close to engineering human capabilities. Hence, the concurrence of science and the everyday-life scenario figures the robotic hand as "almost human" (cf. Castañeda and Suchman 2014). The visually and narratively enacted figure conflates science and everyday life, and, through the jargon entailed, addresses the Hand as being in the making. The combination of the figurative scenario and narratives of making characterizes the hand as belonging to the sphere of technoscientific progress, while equally enacting its distinct difference from typical robotic hands.

4.4.2 *Mirage* in its Exhibition, Online Articles, and Prizes

Recognition of *Mirage* is related to a central passage point, its first exhibition. This event is a material and physical collision of *Mirage*'s technical form, authors, audience, and the discursive practices of media art. From there, further recognition through online articles and media art prizes is unfurled.

Exhibition

LEAP Gallery first exhibited *Mirage* in April 2014 in Berlin. I have already reported on several incidents in the context of this exhibition, as the technical adjustment of *Mirage* to environmental conditions and how Baecker translated his tinkering into a technical drawing. The gallery's name is an acronym for Lab for Emerging Arts and Performance. According to its own description, it is a non-profit interdisciplinary project for emerging art forms, digital media arts, and performance.[60] It hosts international solo and group exhibitions by up-and-coming as well as more established artists. Prior to exhibiting *Mirage*, Baecker had already had a solo exhibition at LEAP Gallery in summer 2013. For that exhibition, Baecker showed several installations, among them *Rechnender*

---

60   LEAP Gallery's website: http://www.xLEAPx.org/ (last accessed August 5, 2014).

Raum. Hence, Baecker and the gallery had already worked together based on their matching artistic interests.

The gallery first exhibited *Mirage* as part of a group exhibition entitled "Obsessive Sensing." In total, the exhibition hosted seven installations from international artists. Most of the installations had been exhibited before. The exhibition flyer introduced the topic of obsessive sensing within a media-philosophical text, in which the author, Sandra Moskova, referred to Max Bense's "information aesthetics." She marked the exhibition's theme as concerned with sensing, computation, and the relations of machines, bodies, materiality, and immateriality. Within this media-philosophical introduction, Moskova mentioned *Mirage* in particular. She referred to *Mirage* as "exploring the borderline between the virtual and the real from two perspectives" (Moskova 2014). On one hand, *Mirage* is an aesthetic investigation of synthesized behavior, and, on the other, the installation investigates the physical space where the virtual becomes real, said Moskova.

The gallery has a main exhibition room of approximately $150m^2$ and an additional room of about $30m^2$. That additional room was dedicated to *Mirage*'s exhibition, as it has no windows and, hence, provides the best light conditions for the laser projection. As I have already reported, Baecker moved to the gallery space several days prior to the opening, as he wanted enough time to adjust *Mirage* to its new environment. However, it was not only *Mirage* that Baecker adjusted. Although the room has no windows and is located in the back of the gallery space, it was still too bright, as the room's entrance has no doors. This caused Baecker to build and install his own curtain instead of doors so as to optimize the light conditions. In so doing, he anticipated how the visitors would recognize the laser reflections on the acrylic foil. Eventually, he placed *Mirage*'s frame so that very little light could fall on its hardware and so that the image projected on the wall was at the darkest end of the room.

The setting prearranged how visitors engage with *Mirage*. At the exhibition opening, I was in the audience. Typically, visitors would come into the room and immediately look at the projection at the dark end of the room. This caused them to pause shortly to perceive the movement. Most visitors go closer to the metal frame. First, they looked at the complete frame and then started to inspect the hardware in more detail. They leaned forward to see the small movements of the wires; some needed to stand on their tiptoes to look at the mirror foil from above; others bent over to look at the small Raspberry Pi display, whose numbers changed frequently; a few visitors even touched the frame, while other visitors weaved through the laser light so as to interrupt the projection. After inspecting the hardware, some visitors moved to the other end of the room to have a closer look at the projected image. Despite this physical inspection, visitors also read the exhibition flyer, which was only possible close to the curtain, or they talked with each other about *Mirage*. Occasionally, Baecker explained how the installation worked and what hardware components he had assembled.

The exhibition's opening was a site where authors, objects, and discourses could collide, which made this discursive practice become a physical and temporal experience. The setting included *Mirage*'s references, which circulated among the audience members in the exhibition flyer, *Mirage*'s physical and aesthetic presence, and in the audience itself, who bodily experienced the laser image and re-enacted *Mirage*'s narratives in discussions about the piece. The temporal collision of all these entities constituted *Mirage* as marked and categorized as an exhibited artwork – its value might not have been exhaustively attributed yet, but the exhibition practice marked *Mirage* as a candidate for valorization through the media art discourse.

Online Articles

The exhibition of *Mirage* at LEAP Gallery was a kick-off for the installation's recognition in online blogs and journals. Some of these publications have been concerned with the group exhibition, while others address *Mirage* in particular. The publishing sites are either explicitly related to art and media art or address current technological trends.

Articles that address the group exhibitions begin by reiterating its theme and explain what "Obsessive Sensing" is about. Siofra McSherry, writing for the contemporary online art magazine "This Is Tomorrow," explained LEAP's new exhibition as "exploring the psychogeography of spaces where digital and physical forms of perception meet." She stressed that "the artists force the viewer to confront the consequences of relying on technology to mediate our experiences" (McSherry 2014). AJ Kiyoizumi added in his article for the online art guide "Berlin Art Link" that the exhibition's concern with technologies is repeated in the practices of the curated artists. He stated that many installations "were created using technical methods rather than traditional artistic-academic skills" (Kiyoizumi 2014). Both authors, McSherry and Kiyoizumi, continued their articles with a run through the exhibited pieces. Their accounts address all of the six pieces, but highlight *Mirage* as a particularly good piece. McSherry marked *Mirage* as "perhaps the most aesthetically successful of the pieces." She connected *Mirage* with the curator's reframing of Max Bense's question of whether aesthetics come after computing. In the following, she sketched *Mirage*'s technical elements and picked up some of Baecker's official tropes, such as the hallucinating algorithm or synthesized landscape. She explained that the "twisting ribbon of red laser projected on the wall is formed from data about the Earth's magnetic field." Furthermore, she added figurative accounts and regarded the grid of muscle wires as resembling "the interior of a piano." Closing the paragraph, McSherry wrote about *Mirage*'s movements (McSherry 2014):

> "The dance between material and immaterial here generates a work of startling and stark beauty, heightened by the impenetrable mysteriousness of the process to the uninitiated."

Kiyoizumi, who regarded the visual aesthetics of *Mirage* as "especially" interesting, also used such figurative language. He attempted to capture these aesthetics in the following sentence (Kiyoizumi 2014):

> "A red wisp seems to tread air on the wall, fluctuating in the same rhythm as a piece of kelp in water."

McSherry's and Kiyoizumi's articles have both been published on websites that are explicitly concerned with current exhibitions in contemporary art and, hence, are generally concerned with the overall exhibition. In contrast, Mitchell Whitelaw's article published on the website "Post Matter" is particularly concerned with *Mirage* and less with the overall exhibition (Whitelaw 2014). Post Matter describes itself as a website that "sits at the convergence of the digital and physical world. Through editorial, exhibitions and art commissions, it celebrates the people and projects that push boundaries."[61] Such self-descriptions are somewhat typical for media art: festivals like Ars Electronica, Transmediale, or Fiber Festival emphasize their boundary positions as, for example, connecting art and science, art and society, technology and culture, etc., or, as here, between the digital and physical. In this sense, Whitelaw's article is located at a more specific site than the previous two: it is not contemporary art in general, but more specifically positioned as part of the media art discussion. Furthermore, Whitelaw is an author engaged in meta-discussion about media and media art. For example, he has published an article about Baecker's work in the refereed journal "Scan – Journal of Media Arts Culture," in which he addressed some of Baecker's previous installations as a novel approach to combining materiality and algorithms (Whitelaw 2013). Hence, Whitelaw did not recognize *Mirage* by accident, as he had been previously engaged with Baecker's work. Furthermore, through his biography as author, his article gains significance as a contribution to the media art discourse.

In this sense, it is no wonder that Whitelaw began his article on Post Matter by referring to Baecker's previous work. This created continuity between *Mirage*

---

61 The Post Matter website: http://postmatter.com/#/home. Self-descriptive quotes are found at the very bottom of the page (last accessed June 26, 2015).

and Baecker's previous installations. Furthermore, it allowed Whitelaw to not only distinguish *Mirage* as a particularly interesting piece, but to also regard Baecker's general artistic approach as different from other media artists. He saw that difference in media artists' common focus on software as medium, whereas Baecker had turned to combining computational algorithms and materiality. For Whitelaw, *Mirage* continues this "poetic approach to hardware," and he described the artwork's approach as "the artist tuning into the Earth itself." He continued by describing *Mirage*'s hardware elements and how they created the moving image. His language picks up Baecker's tropes of dreaming and hallucinating technology. Similarly to the articles by McSherry and Kyioizumi, Whitelaw added figurative accounts to capture the aesthetics of the projected image and described the laser as "creating reflections that play across the room, folding its terrain into a slowly twisting, luminous figure." After describing the image, the article continues with Baecker's general idea behind *Mirage*. Whitelaw reported that Baecker "observes that modern scientific images – from microscopes to telescopes – are increasingly distant from the reality they claim to present." From that observation, Whitelaw enacts in particular *Mirage*'s difference to the scientific machines that hide the connections between the world and the generated image. He formulates *Mirage*'s difference as follows:

> "'Mirage' offers a counterpoint to this: a mechanical image-machine that exposes its inner workings. We can see the muscle wires pulling, the mirror flexing, the laser scanning."

However, Whitelaw has regarded not only the visibility of *Mirage*'s technical components as significant, but also the opacity of how they work together. He described the converting algorithm as "a black box" at the core of the machine and referred to its workings as resembling "a landscape constructed of digital dreams." Closings his article, he reiterated some more of Baecker's figures as a reference to Google data. Significantly, Whitelaw continuously picks up and reformulates figures as "dreaming" or "hallucinating," thereby adding abstract associative descriptions of *Mirage*'s aesthetics to the narratives already written. By enacting *Mirage* as continuing Baecker's work, Whitelaw has signified the installation as a significant contribution to media art. He makes this very explicitly as he enacts differences: firstly, Baecker's approach in opposition to software media artists, and later *Mirage* in opposition to scientific apparatuses. Such differentiating references mark *Mirage* as a contribution to art, in the sense of being different from other art works; furthermore, they mark *Mirage* as a significant contribution to media art, a field that continuously makes cross-links between science and art.

Despite this recognition through art-related publications, *Mirage* is also recognized by websites concerned with technology – or, to put it another way, with modern technological lifestyle. One of such sites is "Wired" magazine, which is a popular print magazine as well as online platform that comments on the latest technology trends. The article by Olivia Solon for Wired shares, with the articles mentioned above, a dense description of how *Mirage* generated its image. In contrast to the articles above, Solon's report also includes interview quotes, in which Baecker speaks about his technical research and about what the installation is supposed to represent. The article emphasizes the technical research behind *Mirage*'s construction. Solon reported on the intense labor and material tinkering that went into *Mirage* and explained how far Baecker's artistic approach involves advanced technologies (Solon 2014). Filip Visnjic, who reported on *Mirage* for "Creative Applications Networks," a website addressing new trends in digital art, puts *Mirage*'s technicity in focus as well (Visnjic 2014). "Amusement," another website concerned with "net culture," started their description of *Mirage*'s technical processes by advising the reader to first "take the time to watch the video and to appreciate the beauty of the landscapes created by the laser projections," before trying to understand the technical explanation, which is a challenging task according to the website (Amusement 2014). The language used in these reports is somewhat less figurative than in Whitelaw's account. The descriptions are more straightforward and do not add associative layers on top of Baecker's own narratives. In particular, Solon's article is concerned with *Mirage*'s technology and less with the artistic concept.

Prizes

As mentioned, media art discourse entails several more or less institutionalized festivals. Some of the more institutionalized festivals include awards for outstanding media art works. Commonly, awards are categorized into subgenres such as visual design, sound art, or the like, which reflect the range of media used in the field. The Prix Ars Electronica, for instance, awards prizes in the categories Computer Animation, Digital Music & Sound Art, Hybrid Art, u19, Visionary Pioneers in Media Art, and the Voestalpine Art and Technology Grant. Each category's main award is the "Golden Nica," which is given to one artist or to one installation. Most categories also entail secondary prizes called "awards of distinction" and "honorary mentions." In order to participate in the Prix, artists need to submit their works according to specific submission guidelines.

Baecker has received several awards, prizes, honorary mentions, and scholarships during his career so far. Among these are second prize at the VIDA 14.0 Art & Artificial Life Award (Madrid, ES), an honorary mention at Prix Ars Electronica

2012, and honorary mentions at Share Prize 2009 (Turin, IT), CYNETart 08 (Dresden, DE), and Digital Sparks Award 08 (Bonn, DE), as well as the production scholarship DOCK e.V. from Schering Stiftung (Berlin, DE). In this regard, the international media art scene has already recognized Baecker's previous installations and, hence, has valuated his work as a significant contribution to the field.

After the exhibition in LEAP Gallery, Baecker submitted *Mirage* to the Prix Ars Electronica 2015 in the category "Hybrid Art." The submission was successful, and *Mirage* received an honorary mention. The Ars Electronica website archives each honorary mention, including several pictures, a video, explanatory notes, and biographical notes on the artist. In total, there are 12 honorary mentions, two awards of distinction, and one Golden Nica in the 2015 Hybrid Art category. The Prix Ars Electronica jury describes the objective of the category as follows:

> "The 'Hybrid Art' category is dedicated specifically to today's hybrid and transdisciplinary projects and approaches to media art. Primary emphasis is on the process of fusing different media and genres into new forms of artistic expression as well as the act of transcending the boundaries between art and research, art and social/political activism, art and pop culture. Jurors will be looking very closely at how dynamically the submitted work defies classification in a single one of the Prix categories of long standing."[62]

This statement stresses boundaries and asks artworks to re-negotiate these in order to receive an award. The jury's statement stresses the fusion of different media that leads to novel forms of artistic expression. According to the jury, not only should this entail the use of different materials, but also the "transcending" of societal realms as art, research, politics, and pop culture. As mentioned, such boundary semantics are typical for media art, and, here, boundaries are enacted so as to serve as criteria for valuable art. Furthermore, the jury's statement not only entails the overcoming of societal boundaries, but also encourages the defying of the classifications of the established Prix Ars Electronica canon, which has been running since 1987. In the latter sense, the jury stresses that, not only is the transdisciplinary character of artworks is matter of value, but also its contribution to media art. This makes the evaluation criteria more specific to the field media art, as it addresses a contribution's relation to previous artworks and that a submission's difference is regarded positively.

The jury's statement indicates that media art and its very criteria reflexively encourage artworks to articulate diverse realms. In so doing, the statement enacts boundaries and renders their overcoming as a value of the media art discourse. Baecker's associative references in his writings and talks mirror such valoriza-

---

62 Prix Ars Electronica website for the category Hybrid Art: http://www.aec.at/prix/en/kategorien/hybrid-art/ (last accessed May 28, 2015).

tion. In this sense, the award values the conceptual and material labor of Baecker's artistic process.

4.4.3 Passage Points, Selected Elements, and Dynamic Practices

The RBO Hand's and *Mirage*'s valorization are both characterized by *passage points* (cf. Callon 1986), which the RBO Hand and *Mirage* have to pass in order to become a recognizable and potentially valuable contributions to robotics and media art. Regarding the RBO Hand, this is the publication of Deimel's and Brock's paper in 2013. All citations refer to the paper and those elements of the RBO Hand that Deimel and Brock have selected and enacted as most relevant. The paper stabilizes what is regarded as relevant to the RBO Hand's approach and, hence, what becomes the central issue in discussions of the Hand's difference and value in contrast to the robotics technologies that are already part of the shared memory. *Mirage*'s passage point is the exhibition at LEAP Gallery. The event was the kick-off for all preceding recognition. It is a public event recognized by authors publishing on contemporary art or media art in particular. The exhibition context connects *Mirage* to a media art institution, LEAP Gallery, and assembles several artworks by different artists under a shared topic of significant concern for the media art discourse. In this regard, these passage points leave their marks on the RBO Hand and *Mirage*: they mark both objects as candidates for valorization, as they have passed the minimum requirement to become part of the shared robotics and media art memory.

The referencing practices of authors discussing the RBO Hand and *Mirage* share references to *selected elements* of both objects. This entails the further signification of those elements that are already focal points of Deimel's and Baecker's referencing practices: compliance and the hallucinating machine. Citations single out the RBO Hand's compliance as the core characteristic of the Hand, and they discuss how far this approach advances robotic grasping and what its impediments are. Furthermore, newspaper articles select and signify the Hand's compliance. They reiterate compliance through tropes like "gentle robot hand" as well as figurative visual codes like placing a coffee cup in the Hand. Concerning *Mirage*, the selective re-enactment of core elements is similar. The exhibition flyer and online articles reiterate Baecker's own writings and add associative references, which further signify the hallucinating machine as *Mirage*'s distinctive concept. The honorary mention by the Prix Ars Electronica mirrors Baecker's concern with scientific technologies and the border between epistemic and aesthetic technologies, embodied in the same trope of the hallucinating machine. In this regard, compliance and the hallucinating machine are recognized and reiterated through the robotics and media art discourse and are further stabilized as categories that embody the difference of both objects.

The valorization of both objects are dynamic *practices*. These practices include markings as well as the production of something else through other authors of a discourse. The RSS Award and the Prix Ars Electronica mark both objects as valuable contributions. Each award includes criteria that emphasize that an object has to go beyond the shared memory of the field – a research paper has to be original and an artwork has to transcend boundaries, so say the criteria. Hence, an award marks an object as distinctively different from all other objects of the shared memory and, in this regard, as a valuable contribution to advancing the field. Marking an object as such is a practice that requires selection, discussion, and statements by jury members. Similar practices are also part of citing and producing articles that refer to the RBO Hand and *Mirage*. Citations and public articles valorize objects, as they are a form of shared sense-making, and discuss to what extent objects embody scientific progress or aesthetic experience. All citations and articles refer to the RBO Hand and *Mirage* in order to produce something more. They use them to legitimize research and illustrate the beneficial aspects of compliant grasping, or they use *Mirage* as an example of media research's concern with boundaries between art and science or the digital and physical realm.

4.5 Articulating Difference as Novelty

In the beginning of this section, I address the objective to delineate the valorization of the RBO Hand and *Mirage*. In order to so, I propose combining elements from Groys' archive figure and Latour's immutable mobiles. Whereas Groys' archive benefits from his clear account of novelty as valorization and the requirement of referring to the old in order to label something as new, his concept lacks dynamics, as he does not account for the selectiveness and materiality of referencing. These impediments are central concerns of Latour's immutable mobiles. They address, in particular, how material representations selectively enfold the world and allow displacement through transformation. The translation of events into objects bridges world and discourse step by step. Nevertheless, Latour's concept lacks what Groys has: a concept of novelty that may account for the valorization of an object, which marks it as a significant contribution to shared memory.

This conceptual discussion structures my empirical account, and I propose regarding the valorization of the RBO Hand and *Mirage* as an articulation of shared discursive practices, translated events, and the referencing practices of authors and archive. Their articulation bridges the laboratory and studio world with the collective valorization of discourses.

4.5 Articulating Difference as Novelty

The *discursive practices* set the scene for both objects' valorization. They are shared, stable, and materialized ways of statement production. In this sense, the RBO Hand and *Mirage* were born into this setting, and discursive practices will continue no matter whether both objects become part of a discourse or not. Institutionalized events are public arenas for objects that allow mutual recognition as well as the capturing of several objects under a shared category, like robotic grasping or hybrid art. Some of such events publish conference proceedings or exhibition catalogues. Publications not only entail written texts about objects, but also distinct and field-specific immutable mobiles like graphs and models, as well as figurative visual codes. Graphs and models are signifiers of science when enacted as conceptual representations. Whereas they are already addressed by Latour, discursive practices in robotics are particularly structured through visual codes that reiterate the humanoid robot as "almost human" (Castañeda and Suchman 2014). Latour's materialistic perspective neglects such shared figures. Nevertheless, they coordinate meaning-making and also signify the difference of a particular research approach in robotics.

The efficacy of discursive practices not only entails public events, but is already demonstrated in the laboratory and studio, as they influence the ways *events are translated* into stable entities. The scientific experiment conducted by Deimel is particularly significant in this regard and does not have a comparative pendant in Baecker's practice. Deimel conducts the experiment according to scripted procedures, which trigger events that either translate into graphs or vanish from experimental results. Discursive practices only allow selected elements of the laboratory world to travel, which either legitimizes certain kinds of failure or conceals illegitimate failure. The experiment significantly enacts the Hand as a scientific object, as it is structured by a sequential order of reproducible differential patterns (cf. Rheinberger 1992) and, furthermore, selectively translates events into scientific representations. The translation of events in Baecker's studio is different in as far as it does not require such a sequential order of events. A drawing enfolds his tinkering so as to connect his future installation and the discursive practices of media art. In contrast to the immutable graphs produced in the robotics experiment, the drawing does not represent *Mirage*'s efficacy, but points to a future enactment of the object. All translations of events share the production of entities that stabilize a temporal object state, select relations while concealing others, and allow for travel beyond the context of their production.

Referencing practices by the authors differ from the translation of events, as far as they do not open opportunities for new relations but actively *add* them. References allude to similarities and differences between the object and other objects, concepts, imaginaries, or figures. Deimel's and Baecker's referencing practices are diverse and entail different media: these range from text-based media, from writings and talks to visual codes in pictures and video to semiotics evoked through the choice of distinct materialities. Despite the variety of media, three kinds of referencing practices dominate the discursive enactment of the RBO Hand and *Mirage*. *Conceptual referencing* dominates Deimel's writings. Compliance is the focal point of references to other robotic hands and concepts, as well as to what the authors regard as the Hand's novelty. Conceptual referencing alludes to similarities and differences between two entities' essential features, which determine their application or technical implementation. In contrast, *figurative referencing* builds upon shared figures of a discourse. References in Deimel's talk reiterate human grasping as an ideal for robotics, which creates stability in shared imaginaries and simultaneously contests previous figurations, as the references allude to specific characteristics, which are enacted as more appropriate, neglected, or different. In contrast, *associative referencing* does not entail such normative statements. In his writings, talks, and online activities, Baecker draws connections between different realms and discourses and alludes to the aesthetic similarities between different scientific technologies. In terms of *Mirage*, associative references allude to technologies' capacity to render otherwise hidden processes, which signifies his focal trope of *Mirage* as a hallucinating machine.

In general, *references through the archive* pick up what authors regard as their objects' distinct features. Archival practices reiterate focal points such as compliance and the hallucinating machine and, in so doing, relate the potentially new to the shared memory of a specific field (cf. Groys 1992). This mutual recognition requires passing *passage points*, which mediate between the potentially new object and the world of discourses. For the RBO Hand, this is the first published paper from 2013, and, for *Mirage*, the passage point is its first exhibition, which kicks off its recognition by journalist, media researchers, and art institutions. Latour's materialistic account of bridging the world and discourse may aid in understanding how immutable mobiles constitute passing through those passage points, but it does not account for the meaning-making via these practices. The paper and exhibition sound a qualitative shift in both objects' existence, as they become recognizable, citable, and discussable through them and, hence, are open to external interpretations and accounts of their value. They become part of shared practices that are concerned with attributing meaning to the object. This particularly shows how authors *select* elements from the paper and exhibition and reiterate these in further publications. This selective reiteration stabilizes the focal points compliance and hallucinating machine. External authors pick them up as categories that embody both objects' differences in contrast to other objects or concepts in the field. Furthermore, they add their own conceptual or associative interpretations to these categories and, hence, signify both categories as meaningful for more general concerns in robotic grasping and

4.5 Articulating Difference as Novelty

media art. This shows that valorization is not just an institutional mark through prizes and awards, which would have been the solitary focus from Groys' perspective. Moreover, valorization is a dynamic practice that re-enacts an object's difference according to one's own research or artistic interests. This is valorization, too, as it fuels practices concerned with inquiries into novelty.

What does the concept of articulation stress in this regard? The main issue of articulation is how a sense of unity evolves through the connection of diverse elements. In this account, I regard discursive practices, translations, and references through authors and archive as articulated. The sense of unity is the focal point that embodies both objects' differences: compliance and the hallucinating machine. All articulated elements increasingly signify these categories from different perspectives and through diverse practices, materialities, and signs. In contrast to Groys, it is not only reference to the old that signifies an object's difference; articulations also stress that difference requires embodiment in order to become a shared value. Such embodiments can be diverse – they can be semantic categories like tropes and visual codes as well as material forms. What they share is the convergence of practices toward increasingly individuated categories that signify the distinct characteristics of an object. In this regard, *difference is not only the relation to what came before, but foremost an articulation of elements into unifying categories that embody an object's value as novelty.*

PART III
Discussion and Conclusions

5 The Aesthetic Reflexivity of Material Practice

The preceding three Chapters delineate results concerning the main question of this study, how do technological objects articulate novelty? The three articulations of novelty – identity, form, and difference – characterize what novelty is in the becoming of technological objects and capture how an object-character as well as an object's difference become part of a shared reality. I wrap up the results in the concluding Chapter 6 and discuss in a condensed way what is seen differently through the lens of articulation in contrast to the established perspectives of inventions, differential pattern, and biographical passage.

For the following Chapter 5, I turn to a question that is somewhat imposed on this study by the empirical findings and, hence, goes beyond its central concerns. The question is, to what extent the RBO Hand and *Mirage* and, in particular, their concern with materiality are significant for new, maybe more reflexive, modes of engaging with technology in science and art? The question aims at discussing in how far the turning back to the more bounded and physical practices of material tinkering in favor of the presumably "emptied-out" (Giddens 1990) abstraction of theory is significant for a specific mode of engaging with technology that can be observed in several other art, design, and science projects as well.

In this sense, the question links back to the intriguing observation that inspired this comparative study in the first place: the poietic engagement with technologically complex objects in science and art. The observation's curiosity was driven by the appearance of technologically complex artworks and the sophisticated engineering they require. Paint and brush, wood and mallet, and the photographic lens seem to be more natural tools and techniques for the production of artworks and the way artists capture and reproduce their life-worlds. Circuits, microchips, algorithms, lasers, and mechanics are typically means for solving concrete problems through engineering and not for aesthetic expression. However, the empirical findings suggest that almost the same could be said concerning the RBO Hand as well, as implementing silicone contradicts the image of high-end scientific technology as well as theory-driven approaches. Plenty, if not most, contributions to robotics are simulations, theoretical models, or based on programming the existing commercial hardware that has become standard infrastructure in robotics research. Such approaches have the benefits that their research is more easily comparable, as they are based on a shared epistemic grammar and do not require additional tinkering skills. Hence, both Deimel and Baecker are not obliged to develop technological hardware in order to contribute to their field, but they still do. The embodiment of the RBO Hand is an approach to creating behavior and effect instead of theoretical modeling – an approach that refers to robotics as an "alchemy," as Brock has said. Baecker's artistic engagement with hardware is a turn toward materiality, too. *Mirage* not only produces an image, but the material and technically complex mode of generating the image also becomes part of the image itself. This deep engagement with materiality was an intriguing observation that inspired this study. It has become a pressing concern through the empirical investigation, which imposes on this study to discuss what signifies such *critical engagement* with materiality and to elaborate whether it is a specific mode of engaging with technology in late modernity.

In so doing, I pick up the notion of "reflexivity" and elaborate its potential to characterize *critical* engagements with technology. Furthermore, I refer to additional examples of science, design, and art projects to discuss if issues learned from comparing the RBO Hand and *Mirage* are of broader significance. For instance, textiles become interfaces, biological probes are perverted as meat stakes, robots perform symptoms of mental illnesses – just a few examples of projects that not only produce an account of or comment on society, but wherein the embodied and material conduct is deeply entwined with epistemologies. I begin the discussion by elaborating different notions of reflexivity in order to differentiate between notions, which regard reflexivity as an elementary part of social conduct and others, which have a more diagnostic connotation so as to characterize shifts occurring in late modernity. From that discussion, I use aspects of the ethnomethodological understanding of essential reflexivity as well as, and more central, diagnostic aspects of aesthetic reflexivity, which is not so much about self-monitoring but foremost about self-*interpretation* (Lash 1993).

5.1 Which Reflexivity?

"Reflexivity" is a term with a diffuse trajectory in social science. It is as widely used as it is differently defined. This means that some definitions understand reflexivity as an anthropological necessity, whereas, in others, the term diagnoses states in the advancement of modern societies. In general terms, Cornelius Schubert distinguishes between three kinds of reflexivity in social-scientific literature (Schubert 2014, 8ff.). He distinguishes between fundamental reflexivity, the reflexivity of social order, and the reflexivity of consequences. *Fundamental reflexivity* refers to the contingency of human actions, in the sense of the capacity for choice and changing the course of effect. This entails the ethnomethodological notion of essential reflexivity, which is concerned with not only how actors are self-aware, but how they make their self-awareness available to others by indicating their rational orientation as part of conducting that particular action (Garfinkel 1967; Lynch 2000; Passoth and Rammert 2016). The *reflexivity of social order* does not contradict this notion, but formulates the question of self-awareness as a matter of social order. This places emphasis on the self-monitoring of societies. Anthony Giddens understands reflexivity in this regard not merely as "self-consciousness," but as the monitored character of the ongoing flow of social life (Giddens 1986). Schubert explains that reflexivity is not merely a matter of doing things differently from that perspective, but includes the maintenance of social order. The *reflexivity of consequences* entails a more diagnostic connotation than the preceding two. This means that reflexivity is used here so as to characterize to what extent modern societies are different from previous societies in terms of the consequences they produce. In Ulrich Beck's diagnosis of a "second modernity," he has understood reflexivity not as increased self-monitoring, but as increasingly unforeseen consequences, unintended effects, and ambiguities in societies of late modernity (Beck 1994).

Another account of reflexivity, which is not covered by Schubert's classification, is *aesthetic reflexivity*, as advanced by Scott Lash and John Urry (Lash 1993; Lash and Urry 1994). They did not distinguish between three types of reflexivity, but rather placed only their aesthetic reflexivity in opposition to cognitive reflexivity. They criticized Giddens and Beck for the hidden cognitivism that underlies their central notions of "monitoring" and "criticism" as drivers of increasingly reflexive modes of social order in late modernity (Lash and Urry 1994, 32-44). According to Lash and Urry, Giddens' and Beck's understanding of reflexivity conceals the aesthetic dimension of reflexivity, which increasingly signifies the expressive dimension of the modern self, whose sources are signs and allegories instead of information and knowledge. This entails, furthermore, a misconception of the body, which is not merely a monitored object, but rather a reflexive agent in itself. In this regard, they propose switching from self-monitoring to self-*interpretation*, which is rooted in hermeneutics (Lash 1993, 8ff.). They have emphasized the interpretative character of processing the sense of information instead of simply feeding information back into practice in the sense of a cybernetic causality. Pivotal concerns of this interpretive flexibility are allegories, bodies, and both's cultural role in late modernity. Lash and Urry's concept is not necessarily an additional type that is missed in Schubert's classification. Rather, it entails elements of ethnomethodology, as both bring individual actors and materiality into the discussion of reflexivity, and, in addition, aesthetic reflexivity stresses social order and consequences, because it regards aesthetics as an integral element of producing order, as well as the diagnosis that aesthetics became increasingly important in the consumption culture at the end of the 20th century.

In the following, I pick up issues from the ethnomethodological basics of reflexivity as well as the diagnostic aspects of Lash's and Urry's work. In so doing, I place emphasis on materiality: firstly, in how references are signified through bodily conduct and then in how materiality becomes an image of critical engagement with science and technology. The section draws on additional examples from robotics and media art in order to delineate reflexive modes of broader significance concerning critical engagements with technology.

5.2 The Indexicality of Referencing

Ethnomethodology understands reflexivity as an essential part of human everyday activity and, as such, as embodied in conduct (Lynch 2000). It is an essential feature of Harold Garfinkel's ethnomethodological program, which is concerned with the everyday accomplishment of organizing activities (Garfinkel 1967). The crux of accomplishing coordinated activities is that actors make available to other actors the rationale of their action *within* the performance of that particular action. The reflexivity of actions is not only the own reaction to oneself within conduct, it is also the "embodied practice through which persons singly and together, retrospectively and prospectively, produce *account-able* states of affairs" (Lynch 2000, 33). Garfinkel explained in this regard that "account-able" means making observable, reportable, and making intelligible one's action, as in making everyday activities recognizable as familiar doings (Garfinkel 1967, 1-9). As such, practice is inevitably indexical. With indexicality, Garfinkel stressed the proclaimed dichotomy between, on one end, abstract or objective statements and, on the other end, those of specific and locally bound meaning. He was radical about this and stated that, in fact, all statements are somewhat indexical, no matter whether it is a scientific text, an administrative file, or a mundane conversation; intelligibility depends on the application of presupposed knowledge in order to characterize the situation at hand and on making sense of statements within that situation (Garfinkel 1967, 18-24). The ability to repair the indexicality of statements in everyday situations lie's in the actor's capacity to act reflexively upon situations, which allows the actor to act in a good-enough manner to continue the flow of conduct.

Although Garfinkel's concerns seem to contradict the study of novelty at first glance, because everyday practice and the accomplishment of organized activity seem to have little to do with articulating difference, his notion of reflexive accountability touches upon issues that are subject matter in this study. Despite the methodological implications that I have already discussed concerning comparison (Stubbe 2015), there is another issue pushed to the forefront via referring to reflexivity in an ethnomethodological sense: *the indexicality of ref-*

5.2 The Indexicality of Referencing

erencing. The articulation of difference entails three modes of referencing: conceptual, figurative, and associative. They are all of a different nature: conceptual referencing alludes to essential features, figurative referencing alludes to desirable figures, and associative referencing alludes to aesthetic similarities. The imposing question in terms of reflexivity is: why are all these different kinds of referencing unproblematic for the actors as well as for the audience? From the present viewpoint, the answer to this is: the indexicality of references is repaired through the actors' performed reflexivity. The sample fingers that Deimel passed around the audience at the ICRA provided an unconventional mode of presenting research at a conference, but still, his actions were intelligible as scientific because they were *performed* as science. Deimel's bodily gestures indicated the connection between the sample fingers and the concepts of his talk. By passing them around in the audience, the silicone's softness became an experience of compliance and, as such, a reference to scientific issues – whereas a change in location and conduct could make the PneuFlex Actuators a toy or whatever else. Baecker's associative referencing is somewhat more drastic in this regard: why are pictures of scientific technologies or geological profiles intelligible as references for art? The performative nature of juxtaposing images without explanations alludes to the aesthetic dimension of scientific technologies. Their aesthetics are not given in the first place, and Baecker does not manipulate or alter the images he finds as scientific illustrations. Rather, the decontextualized reproduction of scientific images makes Baecker's orientation toward their aesthetics available to the audience. His references do not include detailed explanations of models, bullet points, conclusions, or whatever else belongs to presentations in science; Baecker only reproduces images like pictures of apparatuses or graphic representations. As such, the image's intelligibility as an associative reference for art is produced through its decoupling from an original context that still resonates, but that is reduced to visual elements and their peculiar similarities. Deimel and Baecker do not explicitly address or explain the unconventional nature of references; rather, they repair the gap between predictable discursive practices and their own references by making their orientation available to others. In this sense, *not only is referencing a means to create connections to a particular discourse, but its reflexive performance is a signifier of the discourse to which actors contribute*.

The indexicality of referencing focuses on the visual, embodied, and material conduct of making an object's novelty intelligible to others. The reflexivity of this conduct is a matter of accounting for the actor's rationale as well as the significance of material and aesthetic references. However, the ethnomethodological approach to reflexivity might set the basics for reflexivity as a bodily, material, and practical issue, but it does not encourage the differentiation of more or less critical modes of engaging with technology. This concerns how materiality not only is a matter of creating accountability, but itself becomes an image of critical epistemologies.

5.3 Aesthetic Reflexivity and Materiality

Lash and Urry have continued the ethnomethodological concern with individual actors. In contrast to Giddens, they emphasize the interpretive capacity of actors who do not simply seek to secure rules and resources through monitoring, but rather arrange and construe economies of signs as sources for creating a self (Lash and Urry 1994). In so doing, their critique focuses on two issues: bodies and allegories, which they regard as pivotal drivers of aesthetic reflexivity. In the following, I turn to their concern with bodies and draw parallels to material modes of engagement in science and art.

Lash and Urry began with a concern of Giddens, who considers the increasingly individuated body in late modernity. Giddens focuses on how the body turns into an object when actors engage with its outer appearance and monitor its physical state. However, this view reiterates a subject-object dualism, says Lash and Urry's critique, which is deemed inappropriate, as the body itself becomes an agent of reflexivity. They draw on Marcel Mauss to put forward the argument that bodies are not merely mastered by a cognitive self, but rather, the body makes up the conscious and unconscious mind and constitutes practical forms of reasoning. They have cited Mauss in this regard, for whom the body is man's first and most natural instrument (Lash and Urry 1994, 46). In this vein, their understanding of reflexivity gains an experiential notion in which the body constitutes the very process of monitoring and interpretation. This makes self-interpretation an immediate and physically bounded practice, in opposition to the processing of abstract symbols, which "empty out" categories through their mediated forms of engagement. In their argumentation, Lash and Urry have drawn a parallel to the engagement with objects and classifications. They have stressed that the hermeneutic tradition is not engaged in legislating and explaining unmediated universals, but in interpreting and understanding the particularity and groundedness of experience (Lash and Urry 1994, 49). This reasoning is inevitably aesthetic, as it engages senses, feelings, and interpretation.

The affective qualities of reasoning resonate in Deimel's and Baecker's engagement with technology. On the one hand, their practices in the laboratory and studio continuously draw on their bodies as epistemic tools that conduct the constellation of technical infrastructures, forms, and symbols. On the other hand, they employ immediate forms of reasoning through their engagement with technological hardware. Embodiment is not a matter of medium in their practice, in the sense of a concept that is inscribed and transmitted through an artifact. Rather, the material qualities of the artifact are engaged in how concepts come about. The RBO Hand implements silicone to create behavior, which is explicitly in contrast to the abstract modeling of robotic grasping. Similarly, Baecker's first test structure for a hallucinating machine consisted of wood, nuts, bolts, and strings, and he rejected the idea of simply rendering algorithms as graphical representation – a turn toward engineering that he shares with cybernetic pioneer Ross Ashby. This implementation of materialities is reflexive, as it counteracts the linearity of technical progress. In other words, the linearity of progress in robotic grasping can be well continued without ever building a robotic hand, and audio-visual simulation are a valued category in media art. Instead, the RBO Hand and *Mirage* are bound to the particularity and groundedness of hardware, which makes them primitive and equally aesthetic. Their techno-aesthetics are not about monitoring and contemplation, but are experienced through the efficacy of their hardware.

The turn toward the immediate qualities of hardware and the changing stance on technology and knowledge through material engagement is of larger significance to science and art. An institutionalized example is the SymbioticA laboratory at the University of Western Australia. According to their own description, SymbioticA is an artistic laboratory dedicated to research, learning, critique, and hands-on engagement with the life sciences. The laboratory is a fully-equipped research facility that offers artists new means of inquiry and "actively use [of] the tools and technologies of science, not just to comment about them but also to explore their possibilities."[63] Their approach contrasts art *about* science, in the sense of art as a critical representation of what goes on in science. Instead, SymbioticA fosters artworks *from* science, whose critical capacity draws upon the doing and making of quasi-scientific objects. One example is the *Tissue Culture & Art Project*, which explores the use of tissue technologies and different gradients of life through the construction and growth of new classes of semi-living objects. For instance, the artists of the project grew a steak of meat from pre-natal sheep cells harvested as part of medical research on tissue-engineering techniques in utero. The indexicality of the steak's physical existence, as an entity in between nature and artificial laboratory, as well as the artistic conduct in a science laboratory, is repaired through the hermeneutic capacities of actors and audience. This practical engagement is not a matter of monitoring science and feeding that information into an artwork; it is the immediate doing of science *as* art.

Another example is the approach of the Design Research Lab, established in 2010 at Berlin University of The Arts. The Design Research Lab works in "interdisciplinary design research projects that mediate the gap between technologi-

---

[63] SymbioticA website: http://www.symbiotica.uwa.edu.au/ (last accessed October 10, 2015).

cal innovations and people's real needs."[64] One of their projects is concerned with using electronic textiles to develop interfaces between cloth and technology. The developed artifacts combine traditional textile production techniques with electronic functionalities. For instance, a traditional scarf made from wool controls an mp3-player. The engaged human body as well as the materiality of textile, with its haptic, soft, and warm qualities, is the crux of that project. This includes the material asymmetry between the wool and mp3-file and the simplicity of the scarf's shape transporting a certain familiarity. The electronic scarf combines existing artifacts and technologies to create a form of bodily engagement with technology illustrated in pictures on the website that show models wearing and interacting with the scarf. The sense of the project is not the deployment of the scarf as a market ready innovation. Rather, the experimental mode of designing that resonates in its appearance creates its meaning. In this regard, the body and the material textile are involved in performing an alternative mode of human-machine-interaction.

The point I want to make here, in this mode of aesthetic reflexivity, is that technology is *not* a matter of stabilizing societal values; rather, it becomes part of the *destabilization* of scientific knowledge and the technoscientific sense of progress. On the one hand, material resistance takes part in articulating novelty, and, on the other hand, the notion of reflexivity stresses that this is not only a matter of epistemic efficacy but also a matter of creating an image of that efficacy. Ethnomethodologists might argue that reflexivity has always meant that cognition, material, and body are always intertwined, and Lash and Urry, and myself, basically agree with that, but aesthetic reflexivity further stresses that *performing* the disruptive, creative, and critical capacity of engaging with materiality is a specific mode that counteracts devout beliefs in technoscientific representations of knowledge. Science and art are different, in the sense that aesthetic reflexivity might be more dominant in art than in science, but the RBO Hand and its modes of enacting its distinct materialities shows that objects of science are also signified through the affective appeal of experience and efficacy and not only through "emptied-out" abstraction.

5.4 From Symbols to Allegories of Technology

The second driver of Lash and Urry's aesthetic reflexivity is allegories. In claiming that allegories are increasingly important as sources of the self, they have drawn on, in particular, Charles Taylor's account of making the modern identity

---

64 Design Research Lab website: http://www.design-research-lab.org/ (last accessed October 10, 2015).

5.4 From Symbols to Allegories of Technology

(Lash and Urry 1994, 51-54). The central concern of that discussion relates to the difference between symbol and allegory, which has been the subject matter of philosophical debates since the 18th century. Whereas the symbol is a sign in which form and content unite, the allegory is a sign that calls the unity of form and content into question. In their common semiotic meaning, symbols are signs that resemble or directly connect to the denoted object. This connection is maintained through habits or through sets of associations that ensure its particular interpretation. In contrast, allegories separate form and content; they break with the notion of expressive unity (Lash and Urry 1994, 53). Allegories transport meaning, but do not represent an idea or object. In this regard, they require interpretation and re-contextualization to be made sense of. One of the most famous allegories is Plato's Allegory of the Cave. In contrast to Taylor, Lash and Urry have stressed that allegories increasingly fuel late modernity's sense of morality and ethics. This does not mean that symbol and allegory cannot co-exist. The symbol retains an expressivist and romantic tradition, like the call to nature of the Green movement in the 1970s with its utopian symbolism. In contrast, the allegorical mode is more impulsive, anarchistic, and connected to urbanism, globalization, complexity, and heterotopian imaginaries. Haraway's figures entail aspects of both: they stress the tropic qualities of technoscientific imaginaries without claiming that these represent science, which is allegorical, but figures like the OncoMouse are still symbols that are produced as signs of unity between utopian salvation and scientific regimes (Haraway 1997).

Robotics is a field in which figurative accounts follow a dominantly *symbolic mode*. This is demonstrated in the human as an ideal as well as in how robots represent the humanoid imaginary. Concerning the RBO Hand, the several figurative accounts of the human hand are somewhat of a surprise, because the Hand does not have a classical anthropomorphic design. Nevertheless, figurative accounts of the human hand continue from making ideas intelligible to presenting the Hand's concept at a conference, to the public recognition of the Hand in online newspapers. As described, these accounts entail re-figurations, in the sense that specific aspects of the human hand are selected and translated into categories that differ from other categories in the field. However, the accounts are symbolic, as they continuously reiterate the human hand as an ideal. For instance, the story on the RBO Laboratory's website about the unconscious mode of human grasping renders human capacities as a benchmark for how robots should grasp. Similarly, Deimel's figurative references select aspects of human grasping and enact these as desirable and more appropriate for robotic grasping. Accounts of the human hand as ideal dominate robotic grasping even when a technical design follows a minimalist instead of an anthropomorphic approach (cf. Balasubramanian and Santos 2014; Controzzi, Cipriani, and

Carrozza 2014). In this sense, the figures in relation to the RBO Hand remain symbolic as they *unite* the human hand and desired capacities.

In this regard, the symbolic mode of robotics does not claim that robotic and human capacities match, but that the human hand and desired capacities are one. What the human *is* can be figured rather differently in robotics and entail aspects of human embodiment, sociability, and emotions, which are differently reiterated in robotics and artificial intelligence research (Suchman 2007, 226-240). The human-*likeness* or the "almost human," is a signifier of the humanoid imaginary in this regard, as robots are repetitively enacted as technologies in the making whose promises always exceed their actual capacities (Castañeda and Suchman 2014). In this regard, questions of symbols are not so much about how close robot and human are, but about how robots represent human-*likeness*. Suchman has critically denoted, in this regard, that she is less worried about "that robotic visions will be realized [...] than that the discourses and imaginaries that inspire them will retrench, rather than challenge and hold open for contest, received conceptions of humanness" (Suchman 2007, 239). Exemplary of this concern is the public presentation of ASIMO, one of the most famous humanoid robots, developed by Honda.[65] ASIMO holds regular public showcases at a Honda presentation center in Tokyo. The showcase is a scripted performance of ASIMO's capacities. The robot enters the neatly polished stage accompanied by several hostesses. The stage is immaculate, since, apparently, the robot has sensory problems when the environment is too dusty. The choreography involves the performance of ASIMO's walking and running capacities as well as its ability to dance. The robot appears somewhat handicapped, as its movements are unnaturally precise and somewhat stiff and jerky. At the end of the performance, children have the opportunity to get their picture taken with the robot, for which ASIMO makes a "victory" gesture. The point I want to make is that ASIMO embodies the humanoid imaginary not in the sense of equaling human capacities, but as a symbol of technological progress and human-*likeness*, which he performs through somewhat funny conduct. In this sense, there is no gap between the robot and the humanoid imaginary that narrates its significance.

The *allegorical mode* differs in so far as allegories mark precisely the gap between the particular and the general – they diffract form and content. In Giddens' sense, media art is somewhat an institutionalized form of reflexivity, as the field's self-understanding entails commenting on contemporary technological development and feeding statements back into the public discourse about technology (cf. Giddens 1986). Festivals like Ars Electronica are examples of this normative societal self-monitoring through media art. Nevertheless, this form of

---

65 The example is based on my own observations during one of the ASIMO performances in Tokyo, December 2013.

reflexivity does not necessarily mean a move from symbol to allegory, which characterizes aesthetic reflexivity. Allegories point more toward the specific character of references.

This specific, and reflexive, character of allegorical meaning is shown in Baecker's approach to technology. So far, I have addressed Baecker's references as a form of associative referencing that alludes to aesthetic similarities, particularly through sequences of images that evoke a shared imagination of how technologies render hidden processes visible. If one considers his general artistic interest, Baecker's art has a pivotal allegorical element, that is, how he implements technological hardware in order to render visible the technical construction of scientific knowledge. For instance, his installation *Rechnender Raum*, for which Baecker assembled a large wooden frame, a mesh of strings, mechanical pulleys, and electric motors, renders visible the contingency created through increasingly complex connections between technically primitive elements. In a similar vein, *Mirage* is not a representation of the Earth's magnetic field; rather, the complete technical apparatus marks the technical production of scientific images that capture otherwise hidden processes. The figure of a hallucinating machine exaggerates this notion as it counteracts machinic behavior with an unconscious, irrational, and uncontrollable aspect of human life. In this sense, *Mirage* is an installation that marks exactly the technical mediation between the particular (magnetic field) and the general (its image).

Bill Vorn's robotic art is another example of allegoric references, which contrasts the symbolic enactment of humanoids in robotics.[66] In his continuously running project *Robography*, Vorn creates robotic creatures that enter into wild and anarchic forms of interaction. One installation is *DSM-VI*, a robot that expresses "symptoms of 'abnormal' psychological behaviors." The installation's title refers to a scientific document of modern psychiatry: the *Diagnostic and Statistical Manual of Mental Disorders*. Actually, the volume's fifth edition was released in 2013. Hence, the artist proposes a subsequent sixth edition that is concerned with the "the misery of the machines." This project refigures the notion of the human as an ideal by rendering misery and disease instead of embodiment, emotion, and sociability, which are common figures of the human in robotics. In so doing, *DSM-VI* points its finger right at the gap that ASIMO conceals: that between the particular of material technology and the ideal of technoscientific progress.

The meaning of allegories is not intuitive. Whereas ASIMO is fun because kids can recognize it as a humanlike, friendly machine, *Rechnender Raum* and *DSM-VI* are disruptive and require interpretation. Their aesthetic reflexivity is in

---

66 Bill Vorn's website: http://billvorn.concordia.ca/menuall.html (last accessed October 13, 2015).

how their technologies render visible and simultaneously counteract established borders between art and science, human and technology, as well as artifact and imaginary. Technologies are aesthetically reflexive when they render visible the social and technical construction of institutionalized modes of representations and foster critical engagement with a technoscientific sense of progress and innovation. *Novelty, in this regard, is not about indulging in an image of progress, but a matter of experiencing and relating to the heterogeneity of semiotics, literacies, and materialities that are bent, torn apart, whirled, and (re-)articulated through technological objects.*

6 Articulating Novelty

How do technological objects articulate novelty? This is the main question of this study, which I pose in the introductory chapter. The question captures concerns about what novelty actually is in the becoming of technological objects and how diversely novelty becomes part of a shared reality. In this study, I propose the perspective of articulation for inquiring the modes of how novelty and technological objects are entwined. Articulations focus on how heterogeneous elements connect and individuate and, in this sense, they capture the central tension of novelty between individuating and relating technological objects. This central tension between becoming distinct and becoming connected to what is already given characterizes the three articulations of novelty – identity, form, and difference – that I delineated by comparing the RBO Hand and *Mirage*. All three articulations stress that novelty is not difference as such, but that novelty requires coherence among elements as to make an object *distinctly* different.

One central motivation for proposing articulations as heuristic is to give an account of novelty that builds upon critical heuristic resources from social science, but without limiting oneself to the common sociological statement, "novelty is what the actors regard as new." Instead, articulations are a tool for delineating certain moments in an object's becoming, from whence things are different, and materialities do things they have not done before and make sense due to their behavior. Articulations make up a lens that focuses on how diverse realms such as signs, materialities, and bodies are connected so as to create a sense of unity among them. This allows for the delineation of moments when things click in, as well as the critical assessment of practices involved.

I begin the following conclusions by giving a condensed summary of the study's main findings. In so doing, I first sharp up what I consider the central tension of novelty, before I move on and re-capitulate the central aspects of identity, form, and difference as articulations of novelty. Each summary includes an additional remark on what is added to the analytic results through the notion of aesthetic reflexivity. The second section includes a discussion of articulations in contrast to the perspectives of invention, differential pattern, and biographical passage. Closing this study, I provide an outlook on future issues in the study of novelty and technology in contemporary societies.

6.1 Individuating and Relating Objects through Identity, Form, and Difference

In the course of this study, I delineate three articulations of novelty: identity, form, and difference. There are plenty of other possible articulations of the RBO Hand and *Mirage* thinkable, depending on particular connections in their situated enactment. Identity, form, and difference are those articulations that significantly characterize the tension between individuating and relating, which I have signified as the elementary issue concerning novelty. In the following, I first sharp up the two poles of this tension and then sum up how the three articulations solve the tension differently.

6.1.1 Novelty as Individuation and Relation

The abstract answer to the main question of the thesis is: *technological objects articulate novelty through individuated qualities while simultaneously relating these to shared structures of meaning.*

Individuation makes an object a distinct entity that is significant through the coherence of its inner elements. This concerns how a technical object pulls in the elements of its composition so as to create unity in its becoming. Technical forms assemble elements so as to create a causal circularity of effects that was not there before. The clicking-in of components and the transducing energies create sense, because they behave in a distinct, perpetuated, and repeatable sequence of effects. However, individuation is not only a matter of material capacities; it continues in figures and discursive categories. Re-figurations create new narratives, which are significant to the coherence of a story's inner elements. Similarly, references need to pin down the difference of an object and articulate stories, materialities, and structures into unifying categories.

Relating renders an object's difference visible. Novel objects are not given parts of our shared realities like institutions, whose legitimacy is internalized by members of society. In contrast, novel objects need to connect to kin objects and render their difference to become significant. This entails technical relations to existing infrastructures, semiotic relations to shared imaginaries, and bodily relations that act out objects that are not yet realized. All these relations need to render the difference from what came before, whether this appears as a re-assemblage or as a rupture in the flow of things. This relational aspect stresses that novelty may change its character in an object's genesis; it is not a stable property of an object, but a temporal articulation of its individuality and difference.

In this sense, both aspects, individuation and relation, signify identity, form, and difference as three articulations of novelty.

6.1.2 Identity

The first articulation of novelty delineates how objects become part of a shared reality without having yet been fully realized. It captures how an object-*character* evolves and how loose elements connect so as to create a sense of unity.

The articulation begins with an empirical observation and its neglect in science studies: the recurrence of *ideas* as an actor's category. Ideas have vanished from science studies, as the term is associated with historical and philosophical paradigms, from which the turn toward practice that came up with ethnomethodological approaches seeks to differentiate itself. However, speaking of ideas is so repetitive during conversation about an object's meaning that it is difficult to work around them. Instead of stressing what histories ideas hide, I propose asking what ideas in stories and practices connect – and, in this sense, what they *do* to an object.

With the notion of novelty, ideas share the inclusion of histories *and* futures. Stories about an object's origins capture the historical element of ideas. By speaking about ideas, actors mark *selected past events* as relevant to how projects involving new objects come up. One significant element that the stories about the RBO Hand and *Mirage* share is the *deviation* that occurred in the material practice. Brock reported that the idea for the RBO Hand stemmed from experiments on robotic perception that suggested building a more competent hand. Baecker reported that finished installations always deviate from his initial ideas, which drives him to continue his artistic research. In their stories, both actors marked selected experimental and tinkering practices as relevant to their new projects, which eventually became the RBO Hand and *Mirage*. By speaking about ideas, they enact a temporal order that marks what matters and what does not matter for identifying a new object. Nevertheless, this is not just a narrative practice. When visiting places of inquiries, one encounters deviation in material practice leaving traces in the laboratory and studio. Preliminary prototypes and abandoned test structures are selected leftovers that sediment the temporal order of past inquiries.

The future orientation of ideas entails enacting an object's potentials. Figurative stories, material concretization, and situated bodily enactments do just that. In *figurative accounts*, ideas render certain capacities as criteria for what an object is to embody in the future. A story that renders unconscious human grasping as an ideal for robotics narratively connects the RBO Hand to the humanoid imaginary. *Mirage*'s idea is enacted through a dream story that picks up shared imaginaries of artificial intelligence. Both stories share the future objects' relations to *and* contesting of their habitual technoscientific imaginaries. Nevertheless, enacting potentials is not only a story-telling practice. For instance, Baecker reported that ideas often begin with a new mechanical connection, and, from there, stories, tropes, and artifacts come together. Similarly, the Starfish Grabber already embodies the silicone's potential for robotic grasping without being a complete robotic hand. The *situated enactment* of preliminary or prototypical assemblages makes a new object's potentials part of a shared reality. Deimel and Baecker used their bodies to act out not-fully-functioning artifacts and perform what future objects will do. Their selection and bodily enactment of certain aspects is a situated accomplishment based on expectations, interactional resources, and present bodies.

I regard the work that ideas do as building an *object identity*. Analogously to Mead's theory of the self, interaction constitutes identity, which implies self-reflection through the mirroring of actions by the generalized other. In this sense, ideas organize what the generalized other for an object is. They signify an object's biographical trajectory and refer to a generalized collective of kin objects, like robotic hands or artificial neural networks. They embed an object in shared imaginaries of technoscientific belief and mutually give reason to their opposition. Analogously to rituals that an identity has to pass through, situations such as an interview make explicit the distinctive character of an object. In situated enactments, actors anchor an object in biographical trajectories as well as shared imaginaries, which make their opposition meaningful and potentially beneficial for the collective. In those situations, ideas speak with a voice of reason that legitimizes an object's oppositional character by referencing experiential learning and promises regarding shared concerns. For Mead, the self is not a bodiless spirit, but material and bodily presence constitutes the experience and reproduction of collective kinship as well as opposition. In this sense, ideas build an object identity, as they simultaneously individuate and relate an object to a generalized collective of kin objects. They are necessary for articulating coherence in the diversity of materialities, stories, and bodies that make an object.

The generalized character of references differentiates the articulation of an object identity from the latter articulation of difference. Whereas difference articulates selected and precise references, an identity articulates generalized accounts of what the other is. At states where objects are not realized yet, their novelty is not a concern of creating a stable category that embodies their difference. Rather, novelty is a concern of creating a sense of unity that connects histories and futures; this is the general, imprecise, and synthesizing role of ideas.

From the discussion concerning aesthetic reflexivity, one can add to these remarks that performing an object identity is not only a matter of connecting diverse elements so as to create a distinct entity, but moreover that the performance as such makes available the heterogeneity of connections that composite an object. In terms of novelty, this does not only concern the essential indexicality of social conduct; rather, it stresses that indexicality allows for coherence among heterogeneous elements as, for instance, connecting figures from diverse realms with unconventional prototypes, without stabilizing their relations in the sense of a working artefact or definite symbolic meaning. Aesthetic reflexivity stresses that coherence among elements is not the stable relation between symbol and connoted object; rather, coherence implies a margin of indetermination that opens an object up for interpretation.

6.1.3 Form

The second articulation delineates novelty as distinct efficacy. It captures how a technical form evolves and how the form articulates the hybrid constellation of its existence. In contrast to the articulation of an object identity, articulating novelty as form emphasizes meaning-making through material efficacies instead of language and the indication of potentials. Forms create behavior whose sense is a matter of efficacy and not an object's anchoring in shared imaginaries. In other words, novelty of form is things doing things that they have not done before.

A technical form is a functioning and stabilized set of relations whose elements become effective through their sequential order. The technicity of an object is the central element in this regard. For Simondon, technological progress is not in increasing the adaptation of technical forms to an environment, but the concretization of their technical elements. He regarded *concretization* as a convergence of technical parts according to their internal fit. The technical object is not the material artifact, but an ideal allocation that is never fully realized and that pulls together the assembled technical parts. In relation to novelty, this brings about the question, *what articulates a technical form*?

When visiting a robotics laboratory and artist's studio, questions of form are problematic in two regards: there are pre-defined technical formats, and the functioning of a form depends on multiple agents. In the first regard, *technical formats* define what features a form needs to embody in order to become a specific kind of object. These features are part of the form, but, instead of emerging as an object's novelty, they are elements with a continuous trajectory of stabilized technical relations. The RBO Hand is technically required to perform a grasping movement in order to carry the label "robotic hand." The new hand can contest common technical features like sensors or actuation principles, but its technical elements need to be able to capture an item and keep it enclosed – this is the core requirement of robotic hands. *Mirage*'s format is pre-defined less strictly. Still, I assign *Mirage* to the format of cybernetic machines, due to its technical capacity for converting contingent signals from the environment. In this regard, a form's novelty needs to articulate these long-term trajectories as a coherent set that shares and differs from pre-defined formats.

The second regard, which makes questions of form problematic, is made up of the *hybrid constellations* in the laboratory and studio that constitute the objects' technical existence. The robotics laboratory makes available a wide range of technical infrastructure. This infrastructure is comprised of standardized robotics systems into which new components like the RBO Hand have to fit. In his studio, Baecker does not have such high-end infrastructures, but several mechanical and digital hardware components that have the potential to enter the form and become part of a new set of technical relations. In this sense, accounting for hybrid constellations stresses the agencies *through and in* which new forms evolve. On the one hand, constellations impose a range of forms, as specific formats are pre-scribed into them, but, on the other hand, they also enable new forms, as constellations allow for the distribution of agency among different entities. This urges the understanding of questions of form as a matter of characterizing the relation between new form, hybrid constellation, and pre-defined formats.

Comparing the RBO Hand and *Mirage* as technical forms is a challenge in its own right, given their different shapes and functionalities. Nevertheless, technical features can serve as dimensions for comparing how differently or similarly they are embodied. Such comparison needs to take into consideration that technical embodiment is an epistemic decision in the first place. The RBO Hand's as well as *Mirage*'s development contrast objects as models or graphical representations. The embodiment of both objects is a matter of creating behavior and physical efficacy instead of only simulating behavior. In this regard, basic decisions like choice of material and manufacturing process transduce up to kinematic architectures and technical outputs. Whereas the RBO Hand's kinematics are simplistic, which makes the Hand an underactuated system with more degrees of freedom than actuators, *Mirage* would instead qualify as a highly overactuated system, as its actuation and transmission principles add up to a complex assemblage of components. Another pivotal difference is the integration of sensors. The RBO Hand's design is explicitly sensor-free, as its grasping exploits the compliant capacities of silicone, in contrast to *Mirage*, whose basic functionality is the conversion of environmental signals into contingent algorithmic patterns. Despite such unique design decisions, both forms' outputs render their efficacy visible. The RBO Hand's output is a distinct grasping style that uses surfaces for resistance for capturing items, and *Mirage*'s image becomes accountable as a rendering of the Earth's changing magnetic field through its contingent movement.

The novelty of form is an articulation of a form's inner elements and the hybrid constellation of its existence – when parts click in and articulate multiple capacities. Concerning the RBO Hand, this is when the form combines the directed deformability of the silicone fingers with the passive resistance of the palm in such a way that both connect as a material unit, which is mounted on the Meka arm. Said arm drives it into a grasping position from where the fingers can perform a grasp that exploits the material's compliance. From then on, the potential of the PneuFlex Actuators becomes effective as a distinct type of robotic grasping. In contrast to the RBO Hand, which builds upon the capacities of its constellation, *Mirage*'s form incorporates elements of the constellation in which it evolves. This makes the form's output increasingly complex. Within the accumulation of components is a threshold: when the Arduino board converts the sensor signals; when the output signals trigger the muscle wires; when these pull down the springs that are double-taped to the acrylic foil; when all of the 3x16 contact points contingently make the foil move; when the laser light touches the moving foil; when the angle of the laser light is flat enough to create a three-dimensional focused projection. This is novelty because, when these parts work together, the form articulates its complex technical connections and the aesthetics of a moving image. In sum, the RBO Hand builds upon technical functionalities, whereas *Mirage* incorporates functionalities into its form. The former case fosters technical designs to be specific, to be simple, and to distribute technical agency, whereas the latter fosters complexity and the contingency of technical efficacy. In both cases, *novelty is articulated through the crossing of thresholds that makes parts function in a manner that exploits the capacities of their elements*. This process individuates an object, while still being relational, as it happens within hybrid constellations.

The discussion concerning aesthetic reflexivity stresses that novelty of form is not only significant through its physical efficacy, but furthermore through *destabilizing* technoscientific epistemologies. In both cases, material engagement constitutes practical, experiential, and embodied forms of sense-making. The disruptive, creative, and critical capacity of engaging with materiality is a specific mode that counteracts devout beliefs in theories and, hence, links knowledge back to immediate and physically bounded practices of reasoning. The material practices and the unfolding of novel agencies within hybrid constellations oppose knowledge based on processing abstract symbols into "empty out" theoretical grammar. The reasoning of technical forms is inevitably aesthetic, as it engages experience, efficacy, and interpretation.

6.1.4 Difference

The third articulation captures novelty as valorization. It delineates how discourses valorize objects as contributions that go beyond what is already part of cultural memories. Similarly to articulations of identity and form, the main issue of articulating difference is how a sense of unity evolves through the connection of diverse elements. However, instead of articulating generalized accounts and indicating potentials like identities do, *difference* articulates practices, events, and references in a precise and directed way into focal points that embody an object's novelty.

The articulation begins with *discursive practices*, which set the scene for processes of valorization. Discursive practices are shared, stable, and materialized ways of statement production. Institutionalized events like conferences, exhibitions, and festivals are public arenas that allow mutual recognition as well as the capturing of objects under shared categories. Some of such events publish proceedings or catalogues. Publications not only include written text about objects, but also distinct and field-specific immutable mobiles, like graphs and models, as well as figurative visual codes. Graphs and models are signifiers of science when enacted as conceptual representations. In particular, visual codes structure discursive practices in robotics and reiterate the humanoid imaginary.

Discursive practices influence the practice in the laboratory and studio and, in particular, how *events translate* into stable entities. In this sense, discursive practices dominate robotics experiments, in as much as Deimel has conducted experiments with the RBO Hand according to scripted procedures, which trigger events that translate into data clusters and graphs – legitimate scientific representations. The experiment significantly enacts the Hand as a scientific object, as it reproduces a sequential order of reproducible differential patterns. The translation of events in Baecker's studio is different as far as it does not require such a sequential order of events. One drawing enfolds his tinkering and connects his future installation and the discursive practices of media art. In contrast to the immutable graphs produced in the robotics experiment, the drawing does not represent *Mirage*'s efficacy, but points to a future enactment of the object. All translations of events share that they produce entities that stabilize a temporal object state, select relations while concealing others, and allow travel beyond the context of their production.

By *referencing practices by the authors*, both objects enter into the terrain of cultural memories. References actively *add* new relations to an object, as they allude to similarities and differences between the object and other objects, concepts, imaginaries, or figures. Deimel's and Baecker's referencing practices involve different media, which range from writings and talks, visual codes in pictures and videos, to the choice of distinct materialities. Three kinds of referencing practices are significant in this regard. *Conceptual referencing* dominates Deimel's scientific writings. In his writings, Deimel signifies the Hand's distinct grasping style by referring to compliance, which other robotic hands already partly embody but without exploiting its full potential. Conceptual referencing alludes to essential similarities and differences between an object and archived objects. *Figurative referencing* builds upon imaginaries and shared figures of a discourse. These reiterate, for instance, human grasping as an ideal for robotics, while, at the same time, alluding to specific characteristics that are considered

6.1 Individuating and Relating Objects through Identity, Form, and Difference 241

more appropriate, neglected, and better. In contrast, *associative referencing* does not entail such normative statements. In Baecker's writings, talks, and online activities, he draws connections between different realms and discourses and alludes to their aesthetic similarities. In terms of *Mirage*, associative references allude to scientific technologies and their capacity to render otherwise hidden processes visible. This, in return, signifies Baecker's focal trope of *Mirage*: the hallucinating machine.

In general, *references through the archive* do not stress issues different from those provided as distinct features of their objects by the authors. This concerns foremost the increasingly stabilized focal points of compliance and the hallucinating machine. Nevertheless, there are *passage points* that objects need to pass through in order to become recognizable to a discourse as potential contributions. Regarding the RBO Hand, this was the first published paper, and, for *Mirage*, the passage point was its first exhibition, which kicked off its recognition by journalists, media researchers, and art institutions. The paper and exhibition caused a qualitative shift, because, from then on, both objects have become recognizable, citable, discussable, and, hence, open to external interpretations and accounts. External authors have picked up categories like compliance and the hallucinating machine, which both embody the objects' differences in contrast to other objects or concepts in the field, and add their own conceptual or associative interpretations to these categories. This, in return, signifies both categories as meaningful in terms of their potentials regarding shared issues in robotic grasping and media art. In this regard, valorization is not only an institutional mark through prizes and awards, but also a dynamic practice through which an object's difference is re-enacted according to further research and artistic interests.

Difference, in this sense, articulates discursive practices, translated events, and references in a precise and directed way into focal points that embody both objects' novelties: compliance and the hallucinating machine. All articulated elements increasingly signify these categories through diverse practices, materialities, and symbols. In this regard, *difference is not only the relation to what came before, but foremost an articulation of elements into unifying categories that embody an object's value as novelty.*

Concerning aesthetic reflexivity, the articulation of difference shows that the destabilization of scientific representations through forms, as postulated above, is limited. Scientific representations as, for instance, graphs and models still dominate and structure experimental procedures. However, the diagnostic notion of aesthetic reflexivity entails that different modes of referencing can mix. Certainly, the performance of references is crucial for signifying them as contributions to a specific discourse, but the aesthetic appeal and the experience of haptic qualities makes other than established modes of representation possible. Science and art are different, in the sense that aesthetic reflexivity might be more dominant in art than in science, but the RBO Hand and its modes of enacting its distinct materialities shows that objects of science are also signified through the affective appeal of experience and efficacy and not only through "emptied-out" abstraction.

6.2 Articulations: (Differently) Connected Figures, Technicity, and Enactments

In the following, I discuss articulations as heuristic in contrast to inventions, differential patterns, and biographical passages as approaches to the study of novelty and technological objects. Each perspective contains lessons learned from studying novelty and technological objects, and I re-articulate several of their concepts for delineating identity, form, and difference. However, each of them individually is limited. Their main limit is the missing consideration of how differently an object is articulated in its becoming. Furthermore, the perspectives are stuck either in structural accounts of technological trajectories *or* in the situational production of differential patterns, but do not consider the continuous mediation between structures and practice.

In the beginning of this study, I outline that articulations are not a social theory in the classical sense, as the term does not imply *per se* a pre-assumption about how people act or what the elementary parts of social order are. Rather, articulations make up a heuristic device that draws attention to the ways elements are connected and the sense of unity that evolves at a certain point in their connection. In this sense, they are an *object*-heuristic, as they focus on how coherence evolves, but without pre-defining all elements of an object's composition. It is rather a matter of choosing theories in accordance with one's subject matter and using these to sharpen the focus on what is articulated. In this study, I pre-set three elements in accordance with my interest in the relation between novelty and technology: figures, technicity, and enactments. *Figures* draw attention to the symbolic structures of technoscientific imaginaries, *technicity* emphasizes the convergence of technical elements into units of circular causality, and *enactments* stress the situatedness of rendering objects as part of a shared reality. All those elements are concerned with the tension between individuating and relating an object, which signifies novelty. I choose these elements so as to advance the conceptual critique regarding novelty as invention, differential pattern, and biographical passage.

Before I address the points of critique and the central concerns of articulations more precisely, it is necessary to stress that an abstract heuristic is an approach in contrast to traditional ethnographies. Classical ethnographies would refuse to pre-define what is articulated in favor of finding those elements by observing the actors alone. Refusing a traditional understanding of ethnography is, on the one hand, a matter of linking results back to more general discussions about novelty and technology, which I do in the sections for each respective articulation; on the other hand, and maybe more fundamentally, it is a matter of comparison. Comparing two empirically distant cases that do not recognize each other requires abstract concepts in order to draw connections between them. Abstractness allows connection, because concepts as figures, technicity, and enactments are not context-bound in the first place; rather, how they appear or how they are articulated is a matter of context, location, and time. In this regard, concepts make connections possible and draw attention to how similarly or differently phenomena occur. This implies, for instance, the re-articulation of concepts from inventions, differential patterns, and biographical passages despite my general critique. Reflecting upon those concepts is a matter of making explicit the kinds of connections these concepts make possible (Strathern 1991, 51).

I make the criticism that novelty as *invention* either regards technologies as creations of an ingenious mind, as in the colloquial understanding of invention, or may only account for an invention's trajectory in technical documentation, as done by evolutionary perspectives (Gilfillan 1952; Hughes 1987). Both understandings fall short, as they do not consider the situated diversity of the realms of an object's becoming. In contrast, for articulations, this is a key concern. Articulations stress how elements from diverse realms, such as signs, materialities, and bodies, connect so as to render an object's difference visible. For instance, an object identity connects histories, material prototypes, and shared imaginaries, which makes the object part of a shared reality and, furthermore, makes the object distinct in opposition to a generalized account of kin objects, without being anywhere close to stabilization, diffusion, or full technical realization. In this regard, articulations focus on the blind spot between inventions as creations of the mind, because they take structures like figures and technicity into account, and inventions as stable technologies, because they do not limit an object to the historical documents of a device.

The perspective of novelty as *differential patterns* places emphasis on experimental practice. It is concerned with how interactions of heterogeneous agents, like investigated probes, technical infrastructure, the bodies of scientists, etc., interact and eventually create events that are translated into novel objects (Latour and Woolgar 1986; Pickering 1995; Rammert 1999). However, the perspective is limited by its narrow focus on experimental practice and the inscription of discourses into technical settings. This obscures the power of language and the continuity of long-term semiotic structures like technoscientific imaginaries. Instead, articulations stress the connection of both: materiality *and* structures of meaning. For instance, the articulation of difference shows how long-term imaginaries, which are shared and stable structures of a community, are reiterated and simultaneously contested so as to enact difference as valuable to a discourse. In this regard, articulations are moments of unity that connect long-term structures *and* experimental practice.

The perspective of *biographical passages* takes exactly those long-term processes of changing meaning into account. Biographies are concerned with how the meaning of an object changes depending on its historical context and the community that engages with it (Kopytoff 1986; Groys 1992; Daston 2000; Kubler 2008). However, object biographies are limited, as they do not account for the small-scale *processes and practices* of selecting what becomes part of a biography and how objects translate into other entities so as to make them recognizable as novelty. Furthermore, they do not consider the material manipulations that co-produce different meanings. In contrast, articulations consider the active performing of accounts. For instance, considering an object identity as an articulation places emphasis on the contingency of producing something coherent. It acknowledges the deviation of material practice as well as the situated rendering of an object's difference. In this sense, articulations entail the contingency of small-scale material and embodied interactions, instead of only retrospectively sketching a trajectory of changing meaning.

So what are the central concerns of articulations for studying novelty?

1. Articulations stress that objects exist in *differing modes*, depending on their enactment and the nature of the connections. Each articulation – identity, form, and difference – renders the RBO Hand's and *Mirage*'s novelty visible – but each articulation does so differently and makes different connections. On some occasions, they connect stories, figures, and prototypes so as to signify a not-yet-realized object, while, at other times, articulations connect different agents of a technical constellation to create a distinct kind of technical efficacy. In this sense, an object's embodiment depends on its situated enactment and on whether the connections are of a more semiotic or material nature. However, stressing that connections are different and change over the course of an object's becoming does *not* mean they are ever solely semiotic or material. On the contrary, articulations emphasize that it is always a triad of signs, materialities, and bodies that constitutes technological objects. There is never one without the other. What changes is the different emphasis of elements in how they render an object's difference visible.

2. Another concern of articulations is that novelty is not difference as such, but that difference always requires *coherence*. There needs to be a sense of unity among the assembled elements that make an object. This unity is not a matter of fixing an object's meaning or purpose for all time, as things can always connect differently. In this sense, there is no natural belongingness of elements (cf. Hall 1986). However, in order for loose things to become accountable as an object, its elements have to fit together. Fitting together involves material tinkering as well as storytelling practices. In terms of novelty, this suggests that elements might

belong together because of their similarities as well as because of their differences. It is a matter of creating connections between elements, no matter how complex they are, in order to make objects distinguishable entities.

3a. In terms of technologies, connections between *structure and practice* characterize the complexity of articulations. Structures like *imaginaries and figures* constitute the symbolic habitat for technologies. Actors need to respond to shared symbols, such as humanoid visions, artificial intelligence, or human enhancement, since these narrate the significance of technical objects. The opportunity to contest figures and to propose (reflexively) alternate versions of imaginaries is exactly *because* of their nature as structure. Humanoids, artificial intelligence, energy networks, human enhancement, etc. are all structures of shared meanings that are stable and ubiquitous across societal fields and technical applications. They are not bound to a certain field, but are reiterated in public and commercial debates just as in scientific discourses. The stability of technoscientific imaginaries allows for contesting, re-figuring, or simply connecting them to something they have not been connected to before. They make the soft hand and the hallucinating machine evocative figures that make sense because they are connected to symbolic structures *while also* different from them. In this regard, the significance of articulating difference relies on shared imaginaries as contrasting and contested symbolic structures.

3b. Articulating structure and practice is not limited to symbols; rather, it continues in how *infrastructural and experimental technologies* are related. This does not mean technologies need to adapt to given infrastructures in the sense of a new piece of technology having to link with a functional chain of already implemented apparatuses. On the contrary, novel technological objects need to *articulate* the functional settings of their existence. That is, either they build upon given infrastructures, which increases the agency of a technical constellation, *or* they incorporate functionalities into their form. The former case fosters technical designs to be specific, simple, and distribute technical agency. For instance, the RBO Hand articulates the simplicity of silicone and robotics infrastructure into a distinct grasping style, or another example is in musical instruments that advance a distinct tonality that builds upon the ability of other instruments to take over rhythm sections within ensembles. In this case, the concretizing of an object's inherent qualities articulates the technical structure of its existence when its implementation changes the technical agency of the whole constellation. A different articulation is the incorporation of a constellation's functionalities. The Arduino board in *Mirage*'s development is an example of such an articulation, as it changed its status several times and eventually became a permanent entity in the transformation from signal to image. Another example is musical instruments that incorporate parts of an ensemble, like synthesizers. This might lead either to cheap imitations or to new musical rhythms, such as the electronic bass drum, which became a trademark for techno music. The incorporation of functionalities articulates structure and practice when an object's form relies on less external resources to work and, furthermore, does not imitate a previously separated technology, but concretizes elements into a new kind of functional circularity.

6.3 Outlook

Where do we go from here? The study's approach and objectives are of an abstract nature, a demand stemming from the objective to learn something about novelty and technological objects beyond field-specific modes of production. This abstractness brings along the opportunity that the approach can advance in various empirical, methodological, and conceptual directions.

Empirically, there are various fields where the relation of technology and novelty matters, but where production processes follow different modes – for instance, medicine, where technology deeply affects social relations and reconfigures diagnostic as well as ethical relations. Articulation could be an entry point to studying how new technologies connect ethical discourses, technical capacities as well as patient and practitioner bodies, so as to create a unified image of promises for salvation. Studying different fields with the approach of articulation allows the delineation of different kinds of novelty, which might follow other value and power structures as those in science and art.

Methodologically, the comparative design that connects two presumably distant cases is a key element of this study. The methodology combines the openness of ethnographic approaches with the objective to learn something beyond an empirical case. For science and technology studies, such methodologies should be key concerns as they allow integrating findings from diverse empirical fields. Science and technology are issues studied in increasingly diverse fields as, for instance, their role in political governance, urban planning, and pedagogical settings. Integrating empirical findings from diverse fields as to learn something for another and presumably different case should not necessarily entail giving up delineating the contextualization and locality of phenomena. On the contrary, comparative methodologies allow for stressing the situatedness of practices while equally applying concepts that enable connections between cases so as to learn about the similarities and differences of enactments. In light of the increasing empirical scope of science and technology studies, the development of comparative methodologies is a key concern as to integrate the diversity of empirical findings.

Conceptually, I want to point toward two issues regarding articulations and reflexivity. Articulations have proven to be a fruitful heuristic for this study. However, the conceptual implications of the term would benefit from deeper clarifications of its bifurcating conceptual roots, as well as from marking its relation to other accounts of structure and action in the classics of social theory. For instance, how does the term relate to objectification, which captures the externalization of subjective meaning and its stabilization into categories of a shared reality (Berger and Luckmann 1967)? Objectification addresses human action and its tendency to create and act upon reality as an objective reality. The social construction of this reality furthermore includes structural elements such as institutionalization and legitimization. In this sense, articulations would benefit from clarification of the border to such kin social theories of actions and structure.

The second issue regards the study's concern with aesthetic reflexivity. The account given here is only a starting point for the discussion. Further inquiry into the material dimensions of reflexivity needs to begin with collecting and comparing more cases from different societal fields in order to diagnose the extent and scope of projects that critically engage with technology. This would lead into a conceptual discussion of what signifies these modes: is it the move from symbol to allegory as proposed by Lash and Urry, or are there further transitions that mark an increase or change in reflexivity? This would entail comparative analysis of other technology projects in science and art, as well as moving into fields with a completely different logic of production as in, for instance, industrial engineering.

In the sense of an outlook, I want to close this study by quoting (once again) Gilbert Simondon, whose philosophy is not only about intellectual contemplation, but foremost signified by a deep, critical, and empirical engagement with the materialities of technological objects (Simondon [1958] 2012a, 9):

"Recognition of the modes of existence of technical objects must be the result of philosophic consideration; what philosophy has to achieve in this respect is analogous to what the abolition of slavery achieved in affirming the worth of the individual human being."

References

Akrich, Madeleine. 1992. "The De-Scription of Technical Objects." In *Shaping Technology / Building Society. Studies in Sociotechnical Change*, edited by Wiebe E. Bijker and John Law, 205–224. Cambridge, Massachusetts; London, England: The MIT Press.
Alač, Morana. 2009. "Moving Android: On Social Robots and Body-in-Interaction." *Social Studies of Science* 39 (4): 491–528.
Amusement. 2014. "'Mirage', a Sophisticated Optical Apparatus Imagines What Machines Dream about." http://www.amusement.net/2014/04/18/mirage-sophisticated-optical-apparatus-imagines-machines-dream/ (last accessed October 23, 2014).
Appadurai, Arjun. 1986. "Introduction: Commodities and the Politics of Value." In *The Social Life of Things. Commodities in Cultural Perspective*, edited by Arjun Appadurai, 3–63. Cambridge University Press.
Ascott, Roy. 1968. "The Cybernetic Stance: My Process and Purpose." *Leonardo* 1 (2): 105–12.
Balasubramanian, Ravi, and Veronica J. Santos, ed. 2014. *The Human Hand as an Inspiration for Robot Hand Development. Springer Tracts in Advanced Robotics Volume 95*. Heidelberg, New York, Dordrecht, London: Springer.
Barad, Karen. 2007. *Meeting the Universe Halfway. Quantum Physics and the Entanglement of Matter and Meaning*. Durham and London: Duke University Press.
Barthélémy, Jean-Hugues. 2012. "Glossary: Fifty Key Terms in the Works of Gilbert Simondon." In *Gilbert Simondon: Being and Technology*, edited by Arne De Boever, Alex Murray, Jon Roffe, and Ashley Woodward, 203–231. Edinburgh: Edinburgh University Press.
Beck, Ulrich. 1994. "The Reinvention of Politics: Towards a Theory of Reflexive Modernization." In *Reflexive Modernization. Politics, Tradition, and Aesthetics in the Modern Social Order*, edited by Ulrich Beck, Anthony Giddens, and Scott Lash. Cambridge: Polity Press.
Becker, Howard S. 1982. *Arts Worlds*. Los Angeles: University of California Press.
Ben Amor, Heni, Ashutosh Saxena, Nicolas Hudson, and Jan Peters. 2014. "Special Issue on Autonomous Grasping and Manipulation." *Autonomous Robots* 36 (1-2): 1–3.
Berger, Peter L., and Thomas Luckmann. 1967. *The Social Construction of Reality. A Treatise in the Sociology of Knowledge*. New York: Anchor Books.
Bicchi, Antonio. 2000. "Hands for Dexterous Manipulation and Robust Grasping: A Difficult Road toward Simplicity." *Robotics and Automation, IEEE Transactions on* 16 (6): 652–662.

Bonilla, M., E. Farnioli, C. Piazza, M. Catalano, G. Grioli, M. Garabini, M. Gabiccini, and Antonio Bicchi. 2014. "Grasping with Soft Hands." In *IEEE-RAS International Conference on Humanoid Robots (HUMANOIDS 2014)*, 581–587. Madrid, Spain.
Bortolon, Liana. 1965. *Leonardo Da Vinci Und Seine Zeit*. Wiesbaden: Emil Vollmer Verlag.
Braun-Thürmann, Holger. 2005. *Innovation. Themen Der Soziologie*. Bielefeld: transcript Verlag.
Brooks, Rodney. 2005. *Menschmaschinen. Wie Uns Zukunftstechnologien Neu Erschaffen*. Frankfurt am Main: Fischer Verlag GmbH.
Butterworth, George, and Shoji Itakura. 1998. "Development of Precision Grips in Chimpanzees." *Developmental Science* 1 (1): 39–43.
Callon, Michel. 1986. "Some Elements of a Sociology of Translation: Domestication of the Scallops and the Fishermen of St. Brieuc Bay." In *Power, Action and Belief: A New Sociology of Knowledge?*, edited by John Law, 196–223. London: Routledge.
Callon, Michel, and Bruno Latour. 1981. "Unscrewing the Big Leviathan: How Actors Macro-Structure Reality and How Sociologists Help Them to Do so." In *Advances in Social Theory and Methodology: Toward an Integration of Micro and Macro-Sociologies*, edited by Karin Knorr-Cetina and A. V Cicourel, 277–303. Boston, London and Henley: Routledge & Kegan Paul Boston.
Cassirer, Ernst. 1985. *Symbol, Technik, Sprache: Aufsätze Aus Den Jahren 1927-1933*. Hamburg: Meiner.
Castañeda, Claudia, and Lucy A Suchman. 2014. "Robot Visions." *Social Studies of Science* 44 (3): 315–341.
Collins, Harry M. 1982. "The Replication of Experiments in Physics." In *Science in Context. Readings in the Sociology of Science*, edited by Barnes. B. and D. Edge, 94–116. Milton Keynes: The Open University Press.
Controzzi, Marco, Christian Cipriani, and Maria Chiara Carrozza. 2014. "Design of Artificial Hands: A Review." In *The Human Hand as an Inspiration for Robot Hand Development. Springer Tracts in Advanced Robotics Volume 95*, edited by Ravi Balasubramanian and Veronica J Santos Editors, 219–246. Heidelberg, New York, Dordrecht, London: Springer.
Dant, Tim. 2001. "Fruitbox/toolbox: Biography and Objects." *AutoBiography* IX: 11–20.
Daston, Lorraine. 2000. "The Coming into Being of Scientific Objects." In *Biographies of Scientific Objects*, edited by Lorraine Daston, 1–14. Chicago, London: The University of Chicago Press.
De Boever, Arne, Alex Murray, and Jon Roffe. 2012. "'Technical Mentality' Revisited: Brian Massumi on Gilbert Simondon." In *Gilbert Simondon: Being and Technology*, edited by Arne De Broever, Alex Murray, Jon Roffe, and Ashley Woodward, 19–36. Edinburgh: Edinburgh University Press.
De Broever, Arne, Alex Murray, Jon Roffe, and Ashley Woodward, ed. 2012. *Gilbert Simondon. Being and Technology*. Edinburgh University Press.
Deimel, Raphael, and Oliver Brock. 2013. "A Compliant Hand Based on a Novel Pneumatic Actuator." In *2013 IEEE International Conference on Robotics and Automation (ICRA) Karlsruhe, Germany, May 6-10*, 2039–2045.

———. 2014. "A Novel Type of Compliant, Underactuated Robotic Hand for Dexterous Grasping." *Proceedings of Robotics: Science and Systems (RSS) Berkeley, CA, July 13-15.*
Dewey, John. 1938. *Logic. The Theory of Inquiry.* New York: Henry Holt and Company.
———. (1934) 1980. *Art as Experience.* New York: Perigee Books.
Dollar, Aaron M., Antonio Bicchi, Mark R. Cutkosky, and Robert D. Howe. 2014. "Special Issue on the Mechanics and Design of Robotic Hands." *The International Journal of Robotics Research* 33 (5): 675–676.
Dollar, Aaron M., and Robert D. Howe. 2010. "The Highly Adaptive SDM Hand: Design and Performance Evaluation." *The International Journal of Robotics Research* 29 (5): 585–597.
Eco, Umberto. 2002. *Einführung in Die Semiotik.* Paderborn: UTB Verlag.
Fleischmann, Monika, and Wolfgang Strauss. 2007. "Interaktive Kunst Als Reflektion Medialer Entwicklung." *Informatik-Spektrum* 31 (1): 12–20.
Fujimura, Joan H. 1987. "Constructing 'Do-Able' Problems in Cancer Research: Articulating Alignment." *Social Studies of Science* 17 (2): 257–293.
Garfinkel, Harold. 1967. *Studies in Ethnomethodology.* Englewood Cliffs, New Jersey: Prentice-Hall Inc.
Giannaccini, M. E., I. Georgilas, I. Horsfield, B. H. P. M. Peiris, A. Lenz, a. G. Pipe, and S. Dogramadzi. 2014. "A Variable Compliance, Soft Gripper." *Autonomous Robots* 36 (1-2): 93–107.
Gibbs-Smith, Charles. 1978. *The Inventions of Leonardo Da Vinci.* New York: Charles Scribner's Sons.
Giddens, Anthony. 1986. *The Constitution of Society. Outline of the Theory of Structuration.* Cambridge: Polity Press.
———. 1990. *The Consequences of Modernity.* London: Polity Press.
Gilfillan, S Colum. 1935. *The Sociology of Invention.* Chicago: Follett Publishing Company.
———. 1952. "Social Implications of Technical Advance." *Current Sociology* 1: 191–207.
Goodwin, Charles. 2000. "Action and Embodiment Within Situated Human Interaction." *Journal of Pragmatics* 32: 1489–1522.
Groys, Boris. 1992. *Über Das Neue. Versuch Einer Kulturökonomie.* München: Carl Hanser Verlag.
———. 1997. "Technik Im Archiv. Die Dämonische Logik Technischer Innovation." In *Technik Und Gesellschaft, Jahrbuch 9*, edited by Werner Rammert and Gotthard Bechmann, 15–32. Frankfurt, New York: Campus Verlag.
Gugutzer, Robert. 2001. "Grenzerfahrungen: Zur Bedeutung von Leib Und Körper Für Die Personale Identität." *Psychologie Und Gesellschaftskritik* 25 (1): 69–102.
Hall, Stuart. 1980. "Race, Articulation and Societies Structured in Dominance." In *Sociological Theories: Race and Colonialism*, edited by Unesco, 305–345. Paris.
———. 1986. "On Postmodernism and Articulation: An Interview with Stuart Hall (edited by L. Grossberg)." *Journal of Communication Inquiry* 10 (2): 45–60.
Haraway, Donna J. 1989. *Primate Visions. Gender, Race, and Nature in the World of Modern Sicence.* New York, London: Routledge.

———. 1991. "A Manifesto for Cyborgs: Science, Technology, and Socialist-Feminism in the Late Twentieth Century." In *Simians, Cyborgs, and Women: The Reinvention of Nature*, 149–181. New York: Routledge.
———. 1997. *Modest_Witness@Second_Millenium.FemaleMan_Meets_OncoMouse*. New York, London: Routledge.
Hayles, N. Katherine. 1999. *How We Became Posthuman. Virtual Bodies in Cybernetics, Literature, and Informatics*. Chicago and London: The University of Chicago Press.
———. 2005. "Computing the Human." *Theory, Culture & Society* 22 (1): 131–151.
Heidegger, Martin. 1977. *The Question Concerning Technology and Other Essays*. New York and London: Garland Publishing, Inc.
Hinton, Geoffrey E, P Dayan, B J Frey, and R M Neal. 1995. "The 'Wake-Sleep' Algorithm for Unsupervised Neural Networks." *Science* 268 (5214): 1158–1161.
Hoel, Aud Sissel, and Iris Tuin. 2012. "The Ontological Force of Technicity: Reading Cassirer and Simondon Diffractively." *Philosophy & Technology* 26 (2) (November 14): 187–202.
Hörl, Erich. 2008. "Die Offene Maschine. Heidegger, Günther Und Simondon über Die Technologische Bedingung." *MLN* 123 (3): 632–655.
Hoskins, Janet. 2006. "Agency, Biography and Objects." In *Handbook of Material Culture*, edited by Christopher Tilley, Webb Keane, Susanne Küchler, Michael Rowlands, and Patricia Spyer, 74–84. Sage Publications.
Hughes, Thomas P. 1987. "The Evolution of Large Technological Systems." In *The Social Construction of Technological Systems. New Directions in the Sociology and History of Technology*, edited by Wiebe E. Bijker, Thomas P. Hughes, and Trevor J. Pinch, 51–82. Cambridge: The MIT Press.
Hutter, Michael, Hubert Knoblauch, Werner Rammert, and Arnold Windeler. 2015. "Innovation Society Today. The Reflexive Creation of Novelty." *Historical Social Research* 40 (3): 30–47.
Iliadis, Andrew. 2013. "Informational Ontology: The Meaning of Gilbert Simondon ' S Concept of Individuation." *Communication +1* 2: 1–18.
Ilievski, Filip, Aaron D. Mazzeo, Robert F. Shepherd, Xin Chen, and George M. Whitesides. 2011. "Soft Robotics for Chemists." *Angewandte Chemie - International Edition* 50 (8): 1890–1895.
Jensen, Casper Bruun, B. H. Smith, G. E. R. Lloyd, M. Holbraad, A. Roepstorff, I. Stengers, H. Verran, et al. 2011. "Introduction: Contexts for a Comparative Relativism." *Common Knowledge* 17 (1): 1–12.
Joas, Hans. 1992. *Die Kreativität Des Handelns*. Frankfurt am Main: Suhrkamp.
Kac, Eduardo. 1997. "Foundation and Development of Robotic Art." *Art Journal* 56 (3): 60–67.
Keller, Reiner. 2007. *Diskursforschung. Eine Einführung Für SozialwissenschaftlerInnen*. Wiesbaden: VS Verlag für Sozialwissenschaften.
Kiyoizumi, AJ. 2014. "Exhibition // Obsessive Sensing at LEAP." *Berlin Art Link*. http://www.berlinartlink.com/2014/04/22/exhibition-obsessive-sensing-at-leap/ (last accessed October 23, 2014).
Knoblauch, Hubert. 2001. "Fokussierte Ethnographie." *Sozialer Sinn* 1: 123–141.

———. 2011. "Alfred Schütz, Die Phantasie Und Das Neue. Überlegungen Zu Einer Theorie Des Kreativen Handelns." In *Die Entdeckung Des Neuen*, edited by Norbert Schröer and Oliver Bidlo, 99–116. Wiesbaden: VS Verlag für Sozialwissenschaften.
Knorr-Cetina, Karin. 1980. "Die Fabrikation von Wissen. Versuch Zu Einem Gesellschaftlich Relativierten Wissensbegriff." *Kölner Zeitschrift Für Soziologie Und Sozialpsychologie, Sonderheft 22*: 226–245.
———. 1988. "Das Naturwissenschaftliche Labor Als Ort Der 'Verdichtung' von Gesellschaft." *Zeitschrift Für Soziologie* 17 (2): 85–101.
———. 2000. *Epistemic Cultures. How the Sciences Make Knowledge*. Cambridge, Massachusetts; London, England: Harvard University Press.
Kopytoff, Igor. 1986. "The Cultural Biography of Things: Commoditization as a Process." In *The Social Life of Things. Commodities in Cultural Perspective*, edited by Arjun Appadurai, 64–91. Cambridge: Cambridge University Press.
Krohn, Wolfgang. 1977. "Die 'Neue Wissenschaft' Der Renaissance." In *Experimentelle Philosophie. Ursprünge Autonomer Wissenschaftsentwicklung*, edited by Gernot Böhme, Wolfgang van den Daele, and Wolfgang Krohn, 13–128. Frankfurt am Main: Suhrkamp Taschenbuch Wissenschaft.
Kubler, George. 2008. *The Shape of Time*. New Haven, London: Yale University Press.
Laclau, Ernesto. 1977. *Politics and Ideology in Marxist Theory: Capitalism, Fascism, Populism*. London: NLB.
Lash, Scott. 1993. "Reflexive Modernization: The Aesthetic Dimension." *Theory, Culture & Society* 10 (1): 1–23.
Lash, Scott, and John Urry. 1994. *Economies of Signs & Space*. London: Sage Publications.
Latour, Bruno. 1987. *Science in Action. How to Follow Scientists and Engineers through Society*. Cambridge, Massachusetts: Harvard University Press.
———. 1992. "Where Are the Missing Masses? The Sociology of a Few Mundane Artifacts." In *Shaping Technology / Building Society*, edited by Wiebe E. Bijker and John Law, 225–58. Cambridge and London: MIT Press.
———. 1999. *Pandora's Hope. Essays on the Reality of Science Studies*. Harvard University Press.
———. 2005. *Reassembling the Social. An Introduction to Actor-Network-Theory*. New York: Oxford University Press Inc.
Latour, Bruno, and Steve Woolgar. 1986. *Laboratory Life. The Construction of Scientific Facts*. 2nd ed. Princeton, New Jersey: Princeton University Press.
Law, John, and Vicky Singleton. 2005. "Object Lessons." *Organization* 12 (3): 331–355.
Lente, Harro Van. 2012. "Navigating Foresight in a Sea of Expectations: Lessons from the Sociology of Expectations." *Technology Analysis & Strategic Management* 24 (8): 769–782.
Lente, Harro Van, and Arie Rip. 1998. "The Rise of Membrane Technology from Rhetorics to Social Reality." *Social Studies of Science* 28 (2): 221–54.
Lischka, Christoph, and Andrea Sick, ed. 2007. *Machines as Agency. Artistic Perspectives*. Bielefeld: Transcript Verlag.
Lynch, Michael. 2000. "Against Reflexivity as an Academic Virtue and Source of Privileged Knowledge." *Theory, Culture & Society* 17 (3): 26–54.

Mackenzie, Adrian. 2001. "The Technicity of Time: From 1.00 Oscillations/sec to 9,192,631,770 Hz." *Time & Society* 10 (2/3): 235–257.
———. 2005. "Problematising the Technological: The Object as Event?" *Social Epistemology* 19 (4) (October): 381–399.
Marx, Karl. 1976. *Capital. A Critique of Political Economy. Vol. 1.* Edited by Frederick Engels. Middlesex: Penguin Books.
Matthes, Joachim. 1992. "The Operation Called „Vergleichen"." *Soziale Welt Sonderband* 8: 73–99.
Mauss, Marcel. 1966. *The Gift. Forms and Functions of Exchange in Archaic Societies.* London: Cohen & West Ltd.
McSherry, Siofra. 2014. "Obsessive Sensing." *This Is Tomorrow.* http://thisistomorrow.info/articles/obsessive-sensing (last accessed October 23, 2014).
Mead, George H. (1934) 1967. *Mind, Self, and Society. From the Standpoint of a Social Behaviorist.* Chicago and London: The University of Chicago Press.
———. 1987. *Gesammelte Aufsätze - Band 2.* Edited by Hans Joas. Frankfurt am Main: Suhrkamp.
Melchiorri, Claudio, and Maketo Kaneko. 2008. "Robot Hands." In *Springer Handbook of Robotics*, edited by Bruno Siciliano and Oussama Khatib, 345–360. Berlin, Heidelberg: Springer.
Mol, Annemarie, and John Law. 1994. "Regions, Networks and Fluids: Anaemia and Social Topology." *Social Studies of Science* 24 (4): 641–671.
Mondada, Lorenza. 2012. "Video Analysis and the Temporality of Inscriptions within Social Interaction: The Case of Architects at Work." *Qualitative Research* 12 (3) (June 6): 304–333.
Morita, Atsuro. 2014. "The Ethnographic Machine: Experimenting with Context and Comparison in Strathernian Ethnography." *Science, Technology & Human Values* 39 (2): 214–235.
Moskova, Sandra. 2014. "Speculative Imagination as 'Psycholiterature' - a Desire for Obsessive Sensing." In *Exhibtion Flyer Obsessive Sensing.* Berlin: LEAP Gallery.
Müller, Kai. 2014. "Regelungstechnik Und Simulation." Hochschule Bremerhaven, Institut Für Automatisierungs- Und Elektrotechnik. Unterlagen Zur Lehrveranstaltung.
Myers, Natasha. 2008. "Molecular Embodiments and the Body-Work of Modeling in Protein Crystallography." *Social Studies of Science* 38 (2) (April 1): 163–199.
Niewöhner, Jörg, and Thomas Scheffer. 2010. "Thickening Comparison: On the Multiple Facets of Comparability." In *Thick Comparison: Reviving the Ethnographic Aspiration*, edited by Thomas Scheffer and Jörg Niewöhner, 1–15. Leiden, Boston: Brill.
Nowotny, Helga. 2008. *Insatiable Curiosity. Innovation in a Fragile Future.* Cambridge, Massachusetts; London, England: MIT Press.
Ogburn, William F. 1964. *On Culture and Social Change. Selected Papers.* Chicago: Phoenix Books.
Passoth, Jan-Hendrik, and Werner Rammert. 2016. "Fragmentale Differenzierung und die Praxis der Innovation: Wie immer mehr Innovationsfelder entstehen." In Innovationsgesellschaft heute: Perspektiven, Felder und Fälle, edited by Werner Rammert, Arnold Windeler, Hubert Knoblauch, and Michael Hutter, 39-68, Wiesbaden: Springer VS.

Penny, Simon. 2000. "Agents as Artworks: And Agent Design as Artistic Practice." In *Human Cognition and Social Agent Technology*, edited by Kerstin Dautenhahn, 395–414. Amsterdam: John Benjamins Publishing Company.

Pickering, Andrew. 1993. "The Mangle of Practice: Agency and Emergence in the Sociology of Science." *American Journal of Sociology* 99 (3): 559–589.

———. 1995. *The Mangle of Practice. Time, Agency and Science*. Chicago, London: The University of Chicago Press.

———. 2002. "Cybernetics and the Mangle: Ashby, Beer and Pask." *Social Studies of Science* 32 (3): 413–437.

———. 2010. *The Cybernetic Brain. Sketches of Another Future*. Chicago and London: The University of Chicago Press.

Pinch, Trevor J., and Wiebe E. Bijker. 1987. "The Social Construction of Facts and Artifacts. Or How the Sociology of Science and the Sociology of Technology Might Benefit Each Other." In *The Social Construction of Technological Systems*, edited by Wiebe E Bijker, Thomas P Hughes, and Trevor J Pinch, 17–50. Cambridge: MIT Press.

Polanyi, Michael. 1962. "The Republic of Science." Edited by E. Shils. *Minerva* 1: 54–73.

Prattichizzo, Domenico, and Jeffrey C. Trinkle. 2008. "Grasping." In *Springer Handbook of Robotics*, edited by Bruno Siciliano and Oussama Khatib, 671–700. Berlin, Heidelberg: Springer.

Rammert, Werner. 1998. "Die Form Der Technik Und Die Differenz Der Medien. Auf Dem Weg Zu Einer Pragmatistischen Techniktheorie." In *Technik Und Sozialtheorie*, edited by Werner Rammert, 293–326. Frankfurt am Main: Campus.

———. 1999. "Weder Festes Faktum Noch Kontingentes Konstrukt: Natur Als Produkt Experimenteller Interaktivität." *Soziale Welt* 50 (3): 281–296.

———. 2008. "Where the Action Is: Distributed Agency between Humans, Machines, and Programs." TUTS-WP-4-2008. Technical University Technology Studies Working Papers. Berlin.

———. 2012. "Distributed Agency and Advanced Technology. Or: How to Analyze Constellations of Collective Inter-Agency." In *Agency without Actors? New Approaches to Collective Action*, edited by Jan-Hendrik Passoth, Birgit Peuker, and Michael Schillmeier, 89–112. London, New York: Routledge.

Reckwitz, Andreas. 2012. *Die Erfindung Der Kreativität. Zum Prozess Gesellschaftlicher Ästhetisierung*. Berlin: Suhrkamp Taschenbuch Wissenschaft.

Reichardt, Jasia, ed. 1968. *Cybernetic Serendipity - The Computer and the Arts*. London: Studio International.

Rheinberger, Hans-Jörg. 1992. "Experiment, Difference, and Writing: I. Tracing Protein Synthesis." *Studies In History and Philosophy of Science* 23 (2): 305–331.

Riskin, Jessica. 2003. "The Defecating Duck, Or, the Ambiguous Origins of Artificial Life." *Critical Inquiry* 29 (Summer): 599–633.

Rogers, Everett M. 2003. *Diffusion of Innovations. Fifth Edition*. New York: Free Press.

Rosenblum, Nancy. 2007. "Chinese Scholars' Rocks." In *Evocative Objects. Things We Think With*, edited by Sherry Turkle, 252.259. Cambridge and London: The MIT Press.

Rosheim, Mark E. 1994. *Robot Evolution. The Development of Anthrobotics*. New York: John Wiley & Sons Inc.
Šabanović, Selma. 2014. "Inventing Japan's 'Robotics Culture': The Repeated Assembly of Science, Technology, and Culture in Social Robotics." *Social Studies of Science* 44 (3) (January): 342–367.
Salvietti, Gionata, Monica Malvezzi, Guido Gioioso, and Domenico Prattichizzo. 2015. "Modeling Compliant Grasps Exploiting Environmental Constraints." In *Proc. IEEE International Conference on Robotics and Automation*. Seattle, USA.
Schmidtchen, Volker. 1997. "Technik Im Übergang Vom Mittelalter Zur Neuzeit Zwischen 1350 Und 1600." In *Propyläen Der Technikgeschichte, Band 2*, edited by Wolfgang König, 209–600. Berlin: Ullstein Buchverlag.
Schubert, Cornelius. 2014. "Social Innovations. Highly Reflexive and Multi-Referential Phenomena of Today' S Innovation Society? A Report on Analytical Concepts and a Social Science Initiative." TUTS-WP-2-2014. Technical University Technology Studies Working Paper. Berlin.
Schulz, S., C. Pylatiuk, and G. Bretthauer. 2001. "A New Ultralight Anthropomorphic Hand." In *Proceedings of the 2001 IEEE International Conference on Robotics 8 Automation, May 21-26, 2001*, 2437–2441. Seoul, Korea.
Schulz-Schaeffer, Ingo. 2013. "Scenarios as Patterns of Orientation in Technology Development and Technology Assessment." *STI Studies* 9 (1): 23–44.
Schumpeter, Joseph A. 1939. *Business Cycles. A Theoretical, Historical and Statistical Analysis of the Capitalist Process*. Edited by Rendigs Fels. New York, Toronto, London: McGraw-Hill Book Company.
———. 2000. "Enterpreneurship as Innovation." In *Entrepeneurship. The Social Science View*, edited by Richard Swedberg, 50–75. Oxford University Press.
Shanken, Edward A. 2002. "Cybernetics and Art: Cultural Convergence in the 1960s." In *From Energy to Information*, edited by Bruce Clarke and Linda Dalrymple Henderson, 155–77. Palo Alto: Stanford University Press.
Simondon, Gilbert. 1958. *Du mode d'existence des objets techniques*. Paris: Aubier.
———. 2009. "The Position of the Problem of Ontogenesis." *Parrhesia* (7): 4–16.
———. (1958) 2012a. *Die Existensweise Technischer Objekte*. Zürich: diaphanes.
———. 2012b. "Technical Mentality." In *Gilbert Simondon: Being and Technologyl*, edited by Arne De Boever, Alex Murray, Jon Roffe, and Ashley Woodward, 1–18. Edinburgh: Edinburgh University Press.
———. 2012c. "On Techno-Aesthetics." *Parrhesia* (14): 1–8.
Slack, Jennifer Daryl. 1996. "The Theory and Method of Articulation In Cultural Studies." In *Stuart Hall: Critical Dialogues in Cultural Studies*, edited by David Morley and Kuan-Hsing Chen, 113–129. London, New York: Routledge.
Slack, Jennifer Daryl, and J. Macgregor Wise. 2005. *Culture + Technology. A Primer*. New York: Peter Lang.
Smith, Wally. 2009. "Theatre of Use: A Frame Analysis of Information Technology Demonstrations." *Social Studies of Science* 39 (3): 449–480.
Solon, Olivia. 2014. "Computer 'Dreams' Projected with Lasers." *Wired*. http://www.wired.co.uk/news/archive/2014-04/17/mirage-dreaming-computers (last accessed October 23, 2014).

Strathern, Marilyn. 1991. *Partial Connections*. Savage, Maryland: Rowman & Littlefield Publishers.

———. 1999. *Property, Substance and Effect. Anthropological Essays on Persons and Things*. London, New Brunswick: The Athlone Press.

Stubbe, Julian. 2015. "Comparative Heuristics from an STS Perspective. Inquiring 'Novelty' in Material Practice." *Historical Social Research* 40 (3): 109–129.

Suchman, Lucy A. 1987. *Plans and Situated Actions. The Problem of Human-Machine Communication*. Cambridge University Press.

———. 2007. *Human-Machine Reconfigurations. Plans and Situated Actions, 2nd Edition*. New York: Cambridge University Press.

———. 2011a. "Subject Objects." *Feminist Theory* 12 (2) (September 7): 119–145.

———. 2011b. "Anthropological Relocations and the Limits of Design." *Annual Review of Anthropology* 40 (1) (October 21): 1–18.

———. 2012. "Configuration." In *Inventive Methods. The Happening of the Social*, edited by Celia Lury and Nina Wakeford, 48–60. New York: Routledge.

Turkle, Sherry. 2007. *Evocative Objects. Things We Think With*. Edited by Sherry Turkle. Cambridge, Massachusetts; London, England: The MIT Press.

Venn, Couze. 2010. "Individuation, Relationality, Affect: Rethinking the Human in Relation to the Living." *Body & Society* 16 (1) (April 9): 129–161.

Visnjic, Filip. 2014. "Mirage – An Optical Projection Apparatus by Ralf Baecker / LEAP." *Creative Applications Network*. http://www.creativeapplications.net/events/mirage-an-optical-projection-apparatus-by-ralf-baecker-leap/ (last accessed October 23, 2014).

White, Lynn. 1962. "The Act of Invention: Causes, Contexts, Continuities and Consequences." *Technology and Culture* 3 (4): 486–500.

———. 1968. "The Invention of the Parachute." *Technology and Culture* 9 (3): 462–467.

Whitelaw, Mitchell. 2013. "Sheer Hardware: Material Computing in the Work of Martin Howse and Ralf Baecker." *Scan* 10 (2).

———. 2014. "Digital Mirage." *Post Matter*. http://postmatter.com/#/currents/digital-mirage (last accessed June 26, 2015).

Wilson, Stephen. 2002. *Information Arts. Intersections of Art, Science, and Technology*. Cambridge and London: MIT Press.

Winnicott, Donald W. 2005. *Playing and Reality*. London, New York: Routledge.

Zielinski, Siegfried. 2011. "Thinking About Art After the Media: Research as Practised Culture of Experiment." In *The Routledge Companion to Research in the Arts*, edited by Michael Biggs and Henrik Karlsson, 293–312. New York: Routledge.

Zisimatos, Agisilaos G., Minas V. Liarokapis, Christoforos I. Mavrogiannis, and Kostas J. Kyriakopoulos. 2014. "Open-Source, Affordable, Modular, Light-Weight, Underactuated Robot Hands." In *IEEE/RSJ International Conference on Intelligent Robots and Systems (IROS)*. Chicago.

Zivanovic, Aleksandar. 2005. "The Development of a Cybernetic Sculptor: Edward Ihnatowicz and the Senster." *Proceedings of the 5th Conference on Creativity & Cognition*: 102–108.